THE SOVIET COSMONAUT TEAM

THE SOVIET COSMONAUT TEAM

A comprehensive guide to the men and women of the Soviet manned space programme.

By Gordon R. Hooper, FBIS

GRH Publications

First published in 1986 by:

GRH Publications,
36, Bury Hill,
Melton 1001,
WOODBRIDGE
Suffolk
IP12 1LF
ENGLAND

British Library Cataloguing in Publication Data

Hooper, Gordon R.
The Soviet cosmonaut team : a comprehensive guide to the men and women of the Soviet manned space programme.
1. Astronauts—Soviet Union—Biography
I. Title
629.45'0092'2 TL789.85.A1

ISBN 0-9511312-0-6

Computer Typesetting by Technical Options, Woodford Green, Essex. Tel. 01-505-3351.

Printed in the United Kingdom by Richard Clay Ltd, Bungay, Suffolk

To Debbie, with all my love.

CONTENTS

Preface

The purpose of this book is to bring together all the available information on the Soviet Cosmonaut Team – its composition, its history, comprehensive biographies of each man and woman, their spaceflight assignments, callsigns, time spent in space – in fact, to present the fullest possible picture of each person. To the author's knowledge, this is the most detailed book of its kind, and it is intended to become the standard reference work on the subject.

The book's appearance has been timed to mark the 25th anniversary of the World's first manned spaceflight by the Soviet cosmonaut, Yuri Gagarin, on 12 April 1961. Every cosmonaut who has flown since then is included, although only brief details of their missions are given as the purpose of this book is to concentrate on the lives and careers of the cosmonauts themselves.

It should be noted that throughout the book, the French, Indian and Syrian cosmonauts are treated in the same manner as those cosmonauts from the Intercosmos countries, although France, India and Syria are not members of the Intercosmos organisation.

Furthermore, there is limited speculation in the actual biographies – as they are intended to be factual. Where speculation does exist, other than in the background sections, it is denoted by the use of *italics* in the text.

List of Abbreviations

ASTP — Apollo-Soyuz Test Project.

CapCom — Capsule Communicator – the cosmonaut in Flight Control Centre who conducts Earth-Space communications.

CNES — Centre Nationale d'Etudes Spatiale.

CPSU — Communist Party of the Soviet Union.

EVA — Extra Vehicular Activity – i.e. spacewalk.

FCC — Flight Control Centre i.e. Kaliningrad.

GDR — German Democratic Republic.

MCC — Mission Control Centre i.e. Houston.

NASA — National Aeronautics and Space Administration.

Acknowledgements

The Soviet manned space programme is full of mysteries and there are many questions which remain unanswered concerning crew assignments, composition of selection groups, and even the purpose of some of the missions flown. In this book, I have attempted to solve some of the jigsaw, but I am only too aware that many pieces are still missing, and that some may not have been pieced together correctly !

That some of the pieces were even found, and that this book ever appeared, much credit must go to Rex Hall, FBIS, a respected and leading expert on the Soviet manned space programme. Rex has shown great patience and a tremendous willing to help with some of the fine detail of this book, and I am greatly indebted to him.

The other person whose support and patience were essential was my wife Debbie. Not only did she willingly learn to use a word-processor to help with the preparation of the draft, but she also gave me her understanding and every encouragement to finish the book. For that I will always be eternally grateful.

Other people who have helped tremendously in the preparation of the book include Ralph F. Gibbons, FBIS, who has supplied me with much Soviet material over the years; Gilbert Gude, Director of the Congressional Research Service of the Library of Congress, who provided copies of CRS reports on the Soviet space programme; Mark Jones of the Novosti Press Agency, who gave invaluable help in providing most of the photographs used in this book; the Press Sections of the Bulgarian, Cuban, Czechoslovakian, German Democratic Republic, Hungarian, Mongolian, Polish, Romanian, Syrian and Vietnamese Embassies in London, who, together with CNES, provided photographs of the Intercosmos back-up cosmonauts; Chris Cornwell of Technical Options, whose technical expertise and encouragement helped enormously in getting the book into print; and finally, all the people at Acorn Computers for producing the excellent BBC microcomputer on which this book was written.

Those readers wishing to keep abreast of developments in the Soviet manned space programme, are advised to join the British Interplanetary Society. The BIS publishes a monthly magazine, 'Spaceflight,' featuring news and detailed articles about manned and unmanned spaceflight. The Society also produces a Journal, a Space Education magazine, and organises talks, film shows and other events. The Executive Secretary can be contacted at 27/29, South Lambeth Road, LONDON, SW8 1SZ and will be pleased to supply membership details upon request.

Gordon R. Hooper
12 April 1986

Part 1: Background Sections

Part 1: Background Sections

This part of the book contains 30 background sections giving amongst other things, details of when each cosmonaut was selected, what their support and back-up assignments were, what spaceflights they made, the time they spent in space, the callsign used, etc., and also analyses the crewing policies for the Intercosmos flights, ASTP, the Salyut programme etc.

1.00: Soviet Manned Spaceflight Log

Introduction

The following table details all spacecraft launched under the Soviet manned spaceflight programme. It includes manned spacecraft, and other spacecraft also involved in the manned programme i.e. Salyut space-stations, unmanned Soyuz, Progress tankers, and Cosmos spacecraft which docked with Salyut stations.

It does not include other development flights carried out under the Cosmos label, or the Zond series.

Duration is shown in days:hours:minutes, and this refers to **crew** duration not spacecraft duration. On several missions, crews exchanged ships once they had docked with the Salyut. For example, the Soyuz 26 crew returned to Earth in the Soyuz 27 ship, while the Soyuz 27 crew returned in the Soyuz 26 ship. This procedure was followed to provide a 'fresh' Soyuz ferry craft for the crew remaining onboard the Salyut.

Soviet Manned Spaceflight Log

Date:-	Spacecraft:-	Crew:-	Duration:-
1961			
12 April	Vostok 1	Gagarin	01:48
6 August	Vostok 2	Titov, G	1:01:18
1962			
11 August	Vostok 3	Nikolayev	3:22:22
12 August	Vostok 4	Popovich	2:22:57
1963			
14 June	Vostok 5	Bykovsky	4:23:06
16 June	Vostok 6	Tereshkova	2:22:50
1964			
12 October	Voskhod 1	Komarov Feoktistov Yegorov	1:00:17
1965			
18 March	Voskhod 2	Belyayev Leonov	1:02:02
1967			
23 April	Soyuz 1	Komarov	1:02:48

1968			
25 October	Soyuz 2 [1]	Unmanned	Unmanned
26 October	Soyuz 3	Beregovoi	3:22:51
1969			
14 January	Soyuz 4	Shatalov	2:23:21
15 January	Soyuz 5	Volynov Yeliseyev [2] Khrunov [2]	3:00:54
11 October	Soyuz 6	Shonin Kubasov	4:22:43
12 October	Soyuz 7	Filipchenko Volkov, V Gorbatko	4:22:40
13 October	Soyuz 8	Shatalov Yeliseyev	4:22:51
1970			
1 June	Soyuz 9	Nikolayev Sevastyanov	17:16:59
1971			
19 April	Salyut 1	Unmanned	Unmanned
23 April	Soyuz 10	Shatalov Yeliseyev Rukavishnikov	1:23:46
6 June	Soyuz 11	Dobrovolsky Volkov, V Patsayev	23:18:22
1973			
3 April	Salyut 2 [3]	Unmanned	Unmanned
11 May	Cosmos 557 [4]	Unmanned	Unmanned
27 September	Soyuz 12	Lazarev Makarov	1:23:16
18 December	Soyuz 13	Klimuk Lebedev	7:20:56
1974			
25 June	Salyut 3	Unmanned	Unmanned

3 July	Soyuz 14	Popovich Artyukhin	15:17:30
26 August	Soyuz 15	Sarafanov Demin	2:00:12
2 December	Soyuz 16	Filipchenko Rukavishnikov	5:22:24
26 December	Salyut 4	Unmanned	Unmanned
1975			
10 January	Soyuz 17	Gubarev Grechko	29:13:20
5 April	April 5 anomaly	Lazarev Makarov	0:00:22
24 May	Soyuz 18	Klimuk Sevastyanov	62:23:20
15 July	Soyuz 19	Leonov Kubasov	5:22:31
17 November	Soyuz 20 [5]	Unmanned	Unmanned
1976			
22 June	Salyut 5	Unmanned	Unmanned
6 July	Soyuz 21	Volynov Zholobov	49:06:23
15 September	Soyuz 22	Bykovsky Aksyonov	7:21:52
14 October	Soyuz 23	Zudov Rozhdestvensky	2:00:07
1977			
7 February	Soyuz 24	Gorbatko Glazkov	17:17:26
29 September	Salyut 6	Unmanned	Unmanned
9 October	Soyuz 25	Kovalyonok Ryumin	2:00:45
10 December	Soyuz 26 [6]	Romanenko Grechko	96:10:00

1978			
10 January	Soyuz 27 [6]	Dzhanibekov Makarov	5:22:59
20 January	Progress 1	Unmanned	Unmanned
2 March	Soyuz 28	Gubarev Remek	7:22:16
15 June	Soyuz 29 [7]	Kovalyonok Ivanchenkov	139:14:48
27 June	Soyuz 30	Klimuk Hermaszewski	7:22:03
7 July	Progress 2	Unmanned	Unmanned
8 August	Progress 3	Unmanned	Unmanned
26 August	Soyuz 31 [7]	Bykovsky Jähn	7:20:49
4 October	Progress 4	Unmanned	Unmanned
1979			
25 February	Soyuz 32 [8]	Lyakhov Ryumin	175:00:36
12 March	Progress 5	Unmanned	Unmanned
10 April	Soyuz 33	Rukavishnikov Ivanov	1:23:01
13 May	Progress 6	Unmanned	Unmanned
6 June	Soyuz 34 [8]	Unmanned	Unmanned
28 June	Progress 7	Unmanned	Unmanned
16 December	Soyuz T-1 [9]	Unmanned	Unmanned
1980			
27 March	Progress 8	Unmanned	Unmanned
9 April	Soyuz 35 [10]	Popov Ryumin	184:20:12
27 April	Progress 9	Unmanned	Unmanned
26 May	Soyuz 36 [10]	Kubasov Farkas	7:20:46

5 June	Soyuz T-2	Malyshev Aksyonov	3:22:19
29 June	Progress 10	Unmanned	Unmanned
23 July	Soyuz 37 [10]	Gorbatko Tuan	7:20:42
18 September	Soyuz 38	Romanenko Tamayo-Méndez	7:20:43
28 September	Progress 11	Unmanned	Unmanned
27 November	Soyuz T-3	Kizim Makarov Strekalov	12:19:08
1981			
24 January	Progress 12	Unmanned	Unmanned
12 March	Soyuz T-4	Kovalyonok Savinykh	74:18:38
22 March	Soyuz 39	Dzhanibekov Gurragcha	7:20:43
25 April	Cosmos 1267 [11]	Unmanned	Unmanned
14 May	Soyuz 40	Popov Prunariu	7:20:41
1982			
19 April	Salyut 7	Unmanned	Unmanned
13 May	Soyuz T-5 [12]	Berezovoi Lebedev	211:09:05
23 May	Progress 13	Unmanned	Unmanned
24 June	Soyuz T-6	Dzhanibekov Ivanchenkov Chrétien	7:21:51
10 July	Progress 14	Unmanned	Unmanned
19 August	Soyuz T-7 [12]	Popov Serebrov Savitskaya	7:21:52
17 September	Progress 15	Unmanned	Unmanned
31 October	Progress 16	Unmanned	Unmanned

1983			
2 March	Cosmos 1443 [13]	Unmanned	Unmanned
20 April	Soyuz T-8	Titov, V Strekalov Serebrov	2:00:18
27 June	Soyuz T-9	Lyakhov Alexandrov	149:10:46
17 August	Progress 17	Unmanned	Unmanned
27 September	Launch Abort	Titov, V Strekalov	not known
20 October	Progress 18	Unmanned	Unmanned
1984			
8 February	Soyuz T-10 [14]	Kizim Solovyov Atkov	236:22:50
21 February	Progress 19	Unmanned	Unmanned
3 April	Soyuz T-11 [14]	Malyshev Strekalov Sharma	7:21:41
15 April	Progress 20	Unmanned	Unmanned
8 May	Progress 21	Unmanned	Unmanned
28 May	Progress 22	Unmanned	Unmanned
17 July	Soyuz T-12	Dzhanibekov Savitskaya Volk	11:19:14
14 August	Progress 23	Unmanned	Unmanned
1985			
6 June	Soyuz T-13 [15]	Dzhanibekov Savinykh [16]	112:03:12
21 June	Progress 24	Unmanned	Unmanned
19 July	Cosmos 1669 [17]	Unmanned	Unmanned
17 September	Soyuz T-14 [18]	Vasyutin Grechko [19] Volkov, A	64:21:52
27 September	Cosmos 1686 [20]	Unmanned	Unmanned

1986

20 February	Mir	Unmanned	Unmanned

Notes

1 Soyuz 2 was unmanned, and was used in rendezvous manoeuvres with Georgi Beregovoi in Soyuz 3.

2 Yeliseyev and Khrunov returned in Soyuz 4, having spent 1:23:39 in space.

3 Salyut 2 did not achieve stability in orbit, began to tumble, and broke up before it could be manned.

4 Cosmos 557 has been identified in the West as a successfully orbited Salyut, although the station suffered a propulsion or command-sequence failure, rendering it useless for human occupation. It rapidly decayed in orbital altitude, and re-entered on 22 May 1973.

5 Soyuz 20 was an unmanned spacecraft which docked with Salyut 4. It carried a range of biological specimens, including turtles, but its main purpose was to test the Soyuz in a powered-down mode during a mission lasting just over 90 days.

6 Romanenko and Grechko returned in Soyuz 27, Dzhanibekov and Makarov in Soyuz 26.

7 Kovalyonok and Ivanchenkov returned in Soyuz 31, Bykovsky and Jähn in Soyuz 29.

8 Lyakhov and Ryumin returned in Soyuz 34, which had been launched unmanned following the failure of Soyuz 33 to dock with Salyut 6.

9 Soyuz T-1 was an unmanned test of the new Soyuz-T spacecraft. It docked with Salyut 6 on 19 December 1979, and returned to Earth on 25 March 1980.

10 Popov and Ryumin returned in Soyuz 37, Kubasov and Farkas in Soyuz 35, and Gorbatko and Tuan in Soyuz 36.

11 Cosmos 1267 was a prototype space-station module. It docked with Salyut 6 on 19 June 1981, and was de-orbited with the station on 29 July 1982.

12 Berezovoi and Lebedev returned in Soyuz T-7, Popov, Serebrov and Savitskaya in Soyuz T-5.

13 Cosmos 1443 was a multi-purpose space-station module, similar to Cosmos 1267. It was used for cargo delivery and return, and as a space-tug. It docked with Salyut 7 from 10 March to 14 August, and a descent module returned 350 kg of material on 23 August.

14 Kizim, Solovyov and Atkov returned in Soyuz T-11, Malyshev, Strekalov and Sharma in Soyuz T-10.

15 Dzhanibekov returned with Grechko in Soyuz T-13, leaving Savinykh with Vasyutin and Volkov onboard Salyut 7 in the first-ever partial exchange of crews.

16 Savinykh returned in Soyuz T-14, having spent 168:03:51 in space.

17 Cosmos 1669 was said to be a prototype free-flying unmanned platform, similar in appearance to a Progress spacecraft. It docked with Salyut 7 from 21 July to 29 August 1985, and performed a normal Progress-type mission. Pre-launch reports had referred to the impending launch of Progress 25, and it is believed that Cosmos 1669 was in fact this spacecraft, having been given the Cosmos title after suffering a failure of some kind. In the event, ground controllers overcame the problem, and the docking with Salyut 7 was completed.

18 Vasyutin and Volkov returned in Soyuz T-14 with Savinykh.

19 Grechko returned in Soyuz T-13 (see 15), having spent 8:21:13 in space.

20 Cosmos 1686 was another space-station module, similar to Cosmos 1267 and Cosmos 1443. It docked with Salyut 7 on 2 October 1985, and according to the Soviets, was not equipped with a return capsule. It was therefore thought to have been a test of a new module for the Mir space-station.

1.01: Selection Groups

Introduction

The Soviet cosmonaut team has been built up over a number of years from different selection groups. These groups usually have definite patterns to them i.e. all pilots, all engineers, or all women, although there have been occasions where a selection has involved a mixture of pilots and engineers.

This section details cosmonaut selection procedures, the selection groups themselves, and provides an analysis of each selection.

It should be noted that the civilian engineers all seem to be senior design engineers or hold responsible jobs in the design bureau. As they have designed and built the spacecraft they intend to fly, they do not have to follow the same basic training programmes as the pilots, and thus are available for assignment almost immediately after selection. After making their spaceflight, they return to the design bureau to help improve the design of current and future spacecraft. Only then do they usually become available for their next assignment.

On some occasions, particularly in 1963, potential cosmonauts were actually invited to apply, rather than being called up before a selection board.

Cosmonaut Selection Procedures

Cosmonaut-candidates are currently selected from groups of volunteers having extensive flying experience, as well as from a variety of other appropriate backgrounds. This includes individuals with advanced degrees in medicine, biology, engineering, and physics. The ages of these individuals usually ranges from 25 to 45 years and now includes both men and women. According to some reports, the Soviets are now training cosmonauts from even younger age groups in cosmonautic space clubs.

Rigid medical evaluation is the primary criteria for the selection of candidates. Based on general medical practices, they undergo a multi-team medical selection process.

The Soviets also emphasize extensive psychological testing, which has taken on even greater importance with the long-duration spaceflights onboard Salyut space-stations.

The medical selection of cosmonaut-candidates is conducted in three stages:-

First Stage

Preliminary screening is performed on an outpatient basis by a battery of medical sub-specialists, representing internal medicine, neurology, opthalmology, ear-nose-throat, endocrinology, and other medical specialities. As in any in depth physical examination, a medical history of the individual is compiled. Based on the clinical and historical data, an evaluation board consisting of the aforementioned sub-specialists then recommend whether the candidate is suitable and should continue in the selection process.

Second Stage

The second stage of the medical selection process requires the individual to undergo extensive clinical evaluations. During this time, an in depth medical history re-evaluation takes place in an attempt to exclude hereditary disorder trends, as well as recurring pathological conditions, particularly of such organ systems as the circulatory, respiratory, urinary, intestinal tract and skeletal system.

The physical examination employs the usual hands-on evaluation as well as the most up-to-date technological processes including both X-ray examinations and endoscopy, when indicated, as well as various forms of electrocardiography. Laboratory tests are very extensive ranging from complete blood cell counts to other haematological evaluations. Detailed blood chemistry is performed, encompassing most recognised medically significant enzyme and immune parameters. An extensive neurogical evaluation includes motor co-ordination, sensory and reflex functions of the nervous system, as well as skull X-rays and electroencephalography.

Complete psychological testing includes many of the recognised analyses and recently developed tests. The Soviets' experiences in space-station operations has mandated the need for significant psychological testing. New trends in space psychology such as psychological support during the flight and selection of compatible crews has gained great favour. During the training process, emphasis is continually placed on psychological evaluation and selection in order to find individuals particularly suited for professional work with specific space equipment.

An otolaryngological examination includes exo and endoscopy, X-rays of the nasal sinuses, as well as auditory and vestibular function. In view of the early recognition by the Soviets of the balance and orientation difficulties encountered during the early stages of a spaceflight, they have placed great emphasis on evaluating the candidate's vestibular system. A more functional examination includes physiological effects of moderate degrees of hypoxia concurrent with recordings of ECG and arterial pressure. This is followed by a 'dive' test to evaluate adequate acclimatisation of the ears and nasal sinuses to changes in barometric pressure. Tests of cardiac function include various parameters of cardiac rhythm during orthostatic (erect posture) and antiorthostatic tests. Physical load studies on the candidates are performed by means of bicycle ergometry. During this period of testing, pulmonary function as well as cystolic and diastolic blood pressure recovery are evaluated. This is followed by numerous electroencephalograms and other instrumentational evaluations of the nervous, circulatory and cardiac system, as well as numerous vestibular stability tests. The ability to withstand centrifugal forces both in the direction of 'head-toe' as well as 'chest-back' are evaluated. The Soviets continue to stress load-testing in the selection of cosmonaut-trainees.

Based on medical and stress testing, the evaluation board either selects the individual for cosmonaut training; postpones selection until certain physiological abnormalities can be treated; or rejects the candidate based on unsatisfactory responses that are not rectifiable over a short time frame.

Third Stage

The final stage of the selection process is conducted at the Yuri Gagarin Cosmonaut Training Centre. A senior medical evaluation commission bases its decision both on physical and psychological results. Selection is then confirmed by the Chief Medical Commission of the Soviet Union, comprising of the leading specialists in clinical and space medicine. All serious psychological illnesses are grounds for rejection. In individual cases, if the candidate possesses certain technical (research) attributes that are required for the mission, efforts are made to treat certain psychological abnormalities. Selection for cosmonaut training may be made following at least one year of complete recovery. Such a selection process, on an individual basis, may also be carried out in cases when the individual cosmonaut-researchers are judged cured of an infectious or non-infectious disease.

In general, individuals are judged unsuitable for cosmonaut training if disorders are found in psychic, nervous, anatomic, or infectious parameters, including venereal, cardiac, ear-nose-throat, opthalmologic, and skin disorders.

Selection Groups

The following table shows the **official** year of selection. Where appropriate, the cosmonaut groups have been sub-divided into engineers and pilots.

1960	Pilots	Belyayev, Bykovsky, Gagarin, Gorbatko, Khrunov, Komarov, Leonov, Nikolayev, Popovich, Shonin, Titov, Volynov, Anatoli, Dmitri, Grigori, Ivan, Mars, Valentin, Valentin 'Dyed', Valentin Junior.
1962	Women	Tereshkova, Tortchillova.
1963	Engineers	Artyukhin, Demin, Zholobov.
	Pilots	Dobrovolsky, Filipchenko, Gubarev, Shatalov.
1964		Beregovoi.[1]
1964	Engineers	Feoktistov, Yegorov.
1965	Engineers	Glazkov, Rozhdestvensky.
	Pilots	Kizim, Klimuk, Sarafanov, Zudov.
1966	Engineers	Anokhin, Grechko, Kubasov, Makarov, Volkov, Yeliseyev, Lazarev.[2]
1967	Engineers	Rukavishnikov, Sevastyanov.
	Pilots	Kovalyonok, Lyakhov, Malyshev.
1969		Patsayev
1970	Engineers	Andreyev, Ivanchenkov.
	Pilots	Berezovoi, Dzhanibekov, Illarionov, Popov, Romanenko.
1972	Engineers	Lebedev.
1973	Engineers	Aksyonov, Ryumin, Strekalov.
1976	Pilots	Titov, Vasyutin, Volkov.
1977	*Doctors ?*	Atkov
1978	Engineers	Alexandrov, Savinykh, Serebrov, Solovyov, Volk.[3]
1980	Women	Savitskaya, *Irina*.
1982 ?	*women ?*	
1984/ 85	*Scientists ?*	*Antosechin*

Notes:

1 see 1963 selection notes.

2 see 1966 selection notes – he was not an engineer like all the other 1966 men.

3 see 1978 selection notes – he was not an engineer like all the other 1978 men.

Intercosmos cosmonauts

The following are the selection groups for the Soviet-Bloc cosmonauts chosen to fly in the Intercosmos programme. They also include the Indian, French and Syrian cosmonauts.

1976	Hermaszewski, Jähn, Jankowski, Köllner, Pelčak, Remek.
1978	Alexandrov, Dediu, Farkas, Ganzorig, Gurragcha, Ivanov, Lopez-Falcon, Magyari, Prunariu, Tamayo-Méndez.
1979	Liem, Tuan.
1980	Baudry, Chrétien
1982	Malhotra, Sharma.
1985	Clervoy, Deshays, Fares, Favier, Habib, Haignere, Patat, Tognini and Viso.[1]

Note:

1 The 1985 group lists the two Syrian and seven French cosmonauts selected that year. At the time of writing, the French cosmonaut assignments for the second joint Soviet-French spaceflight had not been made.

1960 selection

The decision to begin the selection and training of the first group of cosmonauts was taken in 1959. Early that year, a meeting was held of the Soviet Academy of Sciences and those participating were Mstislav Keldysh (Head of Space programmes), Sergei Korolyov (Chief Spacecraft Designer), Lt.General Nikolai Kamanin (Head of the future group of pilot-cosmonauts), Colonel (Medicine) Yevgeny Karpov (Head of the Commission for selecting the candidate cosmonauts) together with other personalities in the Soviet space programme. Following the meeting, a group of aeronautical doctors, placed under the direction of Karpov, started touring Soviet airbases in July 1959, and selected fighter pilots who, having had a mechanical failure during a flight, had returned to base onboard their aircraft rather than ejecting. These pilots were called together at the Bordenko military hospital at the M.V. Frunze Central Airport at Khodynka, Moscow, in October 1959 to undergo a number of tests:-

— a complete medical examination (by oculists, neurologists, laryngologists, dermatologists and stomatologists.)

— psychotechnic tests (by psychologists.)

— special tests to study the behaviour of the inner ear.

— decompression chamber (rising to 12000 m altitude.)

— centrifuge (acceleration up to 10 G.)

Experts screened several hundred candidates, and from a final group of about forty, 20 men were kept back for the final tests. This first group was presented to the Commander-in-Chief of the Air Force, Marshall Verchinine, in January 1960. On 11 January 1960, Pavel Popovich was the first to be officially assigned to the cosmonaut team. Boris Volynov was the second. The selection of the 20 men – all Air Force or Navy pilots – was finalised on 25 February 1960.

However, Georgi Shonin had heard in November 1959 that he was to join the group in March 1960, and Alexei Leonov reported receiving notification of his selection on 5 March 1960.

For a short while, the cosmonaut-candidates lived in a house in Leninsk Prospect, and their place of work was the M.V. Frunze Central Airport. Close at hand was the Zhukovsky Military Air Academy with laboratories, conference rooms and a swimming pool.

The first training session at the M.V. Frunze Central Airport began at 09.00 on 14 March 1960, and this formed their first base until they moved to Zvezdny Gorodok in April 1960. The Cosmonaut Training Centre was formed in September 1960.

Their working day began at 07.00 with a session of gymnastics, and after breakfast at 08.30, they continued working from 09.00 – 12.00. They attended lectures on space-medicine (Professor V.I. Yazdovsky), rocket propulsion (Prof. M.K. Tikhonravov), celestial mechanics (Professor V.I. Sevastyanov), the construction of spacecraft (Professors K.P. Feoktistov and O.G. Makarov), astronomy (a course at Moscow planetarium), and geophysics. Dinner was at 12.30, and the afternoon was devoted to physical training; it consisted of studies of the cosmonauts abilities to withstand the diverse factors of spaceflight. They had to undergo several sessions in the isolation chamber (where they were kept for 7 – 10 days), the thermal chamber (where they endured temperatures of 70° C (158° F) for up to 2 hours), the decompression chamber (where they underwent a simulated altitude of 12000 m), and the centrifuge (where they underwent accelerations of up to 10 G). Vestibular training was conducted in part during gymnastics sessions and also involved sessions in a spacecraft (under the direction of Air Force Colonel M.L. Gallai). They also learned the principals of radio communications, the use of survival equipment at different landing sites, and got used to wearing a spacesuit during their lessons in the centrifuge, catapult-launchings and parachute jumps. They also had to perform a simulated flight for three days in a mock-up spacecraft using a spacesuit and survival systems.

Of the 20 originally selected, only twelve subsequently made a spaceflight. Eight were unsuccessful, and were removed from the cosmonaut programme during training. Only very brief details of these eight have been released, and they have been identified only by their imyas.

The first 20 candidate cosmonauts were:-

Bykovsky, Belyayev, Gagarin, Gorbatko, Khrunov, Komarov, Leonov, Nikolayev,

Popovich, Shonin, Titov, Volynov, Anatoli, Dmitri, Grigori, Ivan, Mars, Valentin, Valentin 'Dyed' and Valentin Junior.

Of the 20, when selected 15 were members of the CPSU and 5 were in the Komsomol.

1962 selection

A group of four women was selected on 16 February 1962, and the first training session was held a month later, on 14 March. The women have officially been identified as Valentina Tereshkova, Irina, Tanya (Tatiana) and an unnamed woman.

All were said to be parachutists but not all had pilot-training. Following the Vostok 6 mission, the three who did not fly returned to their jobs, two as engineers in Star Town, while the other went to work in Moscow.

It now seems that Tanya Tortchillova was the back-up to Tereshkova, (see Tortchillova's biography for full details) although some Western observers had, over a period of years, speculated that the back-up was a woman named 'Vera.' Others had said her name was 'Venera', Russian for Venus, which may have been a codename.

1963 selection

Fifteen men reported for duty at Star Town on 11 January 1963, and reported directly to Pavel Belyayev. So far, only Artyukhin, Demin, Dobrovolsky, Filipchenko, Gubarev, Shatalov and Zholobov have been identified.

They were a more experienced and older group than the majority of the 1960 selection. All had graduated from various institutes, and as well as pilots the new group included military engineers. Of those identified so far, the pilots were Dobrovolsky, Filipchenko, Gubarev and Shatalov, and the engineers were Artyukhin, Demin and Zholobov.

According to one Soviet report, one of the other men selected is still in training, having served as the back-up to Viktor Patsayev on Soyuz 11.

N.B. Beregovoi reported for duty in February 1964, having been specially selected on the personal recommendation of General Kamanin. It was understood for many years that he was a solo 1964 selection, but in 1985, he revealed that he was in fact part of the 1963 selection, and was simply a late arrival. Whether he made the group 16 in number, or was one of the original 15 is unclear.

1964 selection

Feoktistov, Lazarev, Makarov and Yegorov were selected specially for the Voskhod 1 mission. Lazarev and Makarov were subsequently stood down.

N.B. There may have been others in the group – see section entitled 'Voskhod 3 Crewing.'

Grechko, Kubasov, Volkov and Yeliseyev were also interviewed with Konstantin Feoktistov in 1964, but were not called until 1966.

There is some confusion over Yegorov's selection date. Officially, he joined the cosmonaut team in 1964, but there is some evidence to suggest that he was originally selected earlier than this - probably in 1963. (see Yegorov's biography for full details.)

1965 selection

In October/November 1965, the 'Young Guards' group was selected, with 20 men reporting to Star Town. All were young pilots and military engineers whose basic training lasted for two years.

It is believed that the Soviets had decided to see what results could be achieved by choosing an especially young group of men in preparation for the forthcoming Soyuz programme.

Only four pilots – Kizim, Klimuk, Sarafanov and Zudov – and two engineers – Glazkov and Rozhdestvensky – have been identified and it would therefore seem that the failure rate was high.

According to one Soviet report, two engineers are still in training.

1966 selection

In view of the forthcoming first two flights of Soyuz, Soyuz 2 having a crew of a pilot, an onboard engineer, and a research-engineer, a special selection to recruit the onboard engineer was made. 8 flight-engineers were selected on 23 May 1966, but did not report for training immediately. They were all civilian engineers attached to the Korolyov design bureau, and they had all been involved in the design of Vostok, Voskhod and the new Soyuz. This experience enabled them to be available for immediate assignments.

On 5 September 1966, the first 4 arrived in Star Town. They had all been interviewed with Konstantin Feoktistov in 1964, but were not called until 1966. They were Grechko, Kubasov, Volkov and Yeliseyev. Like Feoktistov, it was not necessary for them to complete a full programme of training, and they therefore only spent six months preparing for Soyuz 2.

In November 1966, 3 more engineers arrived in Star Town – Makarov, Rukavishnikov and Sevastyanov – but the latter two did not start training until January 1967.

An 8th man – Sergei Anokhin – joined the team but only stayed a few months before being disqualified due to physical problems.

Lazarev was also selected to supplement the team, having previously acted as back-up with Makarov on Voskhod 1, before being stood down.

1967 selection

On 8 January 1967, Rukavishnikov and Sevastyanov, originally selected in 1966 (see above) reported for duty and became available for immediate assignment.

In May 1967, a group of pilots reported to Belyayev, who was responsible for their initial training. They were probably replacements for pilots removed from the 1965 selection, and the only men identified so far are Kovalyonok, Lyakhov and Malyshev.

1969 selection

Patsayev – another member of the Korolyov design team – was recruited specially to fly onboard the first Salyut space-station.

1970 selection

A new group made up of Air Force pilots and civilian engineers was selected in 1970, and reported directly to Boris Volynov. The pilots – Berezovoi, Dzhanibekov, Illarionov, Popov and Romanenko – followed the previous pattern of being young and inexperienced. The engineers – Andreyev and Ivanchenkov – came from the Korolyov design bureau, and have been described as the best of the young generation of engineers.

Romanenko was selected in 1969, but officially joined the cosmonaut team in 1970. Berezovoi has reported that he was selected in May 1970.

1972 selection

Lebedev – another engineer from the Korolyov design bureau – was a special selection.

1973 selection

Another group of senior design engineers from the Korolyov design bureau joined the cosmonaut team in 1973. Only three have so far been identified – Aksyonov, Ryumin and Strekalov.

According to one Soviet report, another engineer is still in training.

1976 selection

During 1975, it was decided to recruit another Air Force pilot group, and this selection began its basic training in 1976. Only Vasyutin, Volkov and Titov have so far been identified – but it is believed others are still in training. Vasyutin and Volkov both reported for duty in February 1976.

1977 selection

Little is known about the 1977 selection, and the only cosmonaut identified so far is Oleg Atkov. It seems likely that the selection was made up of doctors specialising in space medicine, especially as Beregovoi, when asked if there would be a doctor on the crew of all long-duration flights in the future replied 'Yes, definitely.'

1978 selection

The 1978 selection consisted of engineers drawn mainly from Korolyov's design bureau – Alexandrov, Savinykh, Serebrov and Solovyov – together with a civilian test-pilot – Igor Volk.

Volk and Savinykh were both selected in December 1978.

1980 selection

Four women were selected in 1980. Savitskaya and Irina began training immediately, whereas the other two, as yet unidentified, were not called up until late 1984. Both are apparently researchers, and not pilots or engineers.

1982 ? selection

In an interview with Air and Cosmos on her visit to the IAF in 1984, Savitskaya said that there were 10 women in training as part of the cosmonaut team.

These are probably the four selected in 1980 and a further 6 selected in *1982*?

1984/85 ? selection

In late 1984, Anatoli Antosechin, a Soviet scientist from the Institute of Molecular Genetics said he was about to start cosmonaut training. It is therefore believed that a group of scientists was selected in 1984/85 in preparation for the new space-station expected in 1986.

Intercosmos selections

1976

The first Intercosmos cosmonauts were chosen in December 1976, and consisted of representatives from Czechoslovakia, the GDR and Poland :-

Hermaszewski, Jähn, Jankowski, Köllner, Pelčak, Remek.

1978

The second group of Intercosmos cosmonauts were chosen in March 1978, and consisted of representatives from Bulgaria, Cuba, Hungary, Mongolia and Romania:-

Alexandrov, Dediu, Farkas, Ganzorig, Gurragcha, Ivanov, Lopez-Falcon, Magyari, Prunariu, Tamayo-Méndez.

1979

The third group of Intercosmos cosmonauts were chosen in April 1979, and consisted of the two Vietnamese cosmonauts, Liem and Tuan.

1980

The two French cosmonauts – Baudry and Chrétien – were chosen on 12 June 1980.

1982

The two Indian cosmonauts – Malhotra and Sharma – were chosen in September 1982.

1985

The two Syrian cosmonauts – Fares and Habib – were chosen in late 1985.

In addition, seven French cosmonauts – Clervoy, Deshays, Favier, Haignere, Patat, Tognini and Viso – were chosen in September 1985. At the time of writing, the French cosmonaut assignments for the second Soviet-French spaceflight had not been made.

1.02: Cosmonauts in order of first spaceflight

Introduction

The following table shows each Soviet cosmonaut in order of first spaceflight, using the official Soviet method of determining ranking in the event of two or more cosmonauts making a first spaceflight together: -

1	Yuri Gagarin	31	Gennady Sarafanov
2	Gherman Titov	32	Lev Demin
3	Andrian Nikolayev	33	Alexei Gubarev
4	Pavel Popovich	34	Georgi Grechko
5	Valery Bykovsky	35	Vitaly Zholobov
6	Valentina Tereshkova	36	Vladimir Aksyonov
7	Vladimir Komarov	37	Vyacheslav Zudov
8	Konstantin Feoktistov	38	Valery Rozhdestvensky
9	Boris Yegorov	39	Yuri Glazkov
10	Pavel Belyayev	40	Vladimir Kovalyonok
11	Alexei Leonov	41	Valery Ryumin
12	Georgi Beregovoi	42	Yuri Romanenko
13	Vladimir Shatalov	43	Vladimir Dzhanibekov
14	Boris Volynov	44	Alexander Ivanchenkov
15	Alexei Yeliseyev	45	Vladimir Lyakhov
16	Yevgeny Khrunov	46	Leonid Popov
17	Georgi Shonin	47	Yuri Malyshev
18	Valery Kubasov	48	Leonid Kizim
19	Anatoli Filipchenko	49	Gennady Strekalov
20	Vladislav Volkov	50	Viktor Savinykh
21	Viktor Gorbatko	51	Anatoly Berezovoi
22	Vitaly Sevastyanov	52	Alexander Serebrov
23	Nikolai Rukavishnikov	53	Svetlana Savitskaya
24	Georgi Dobrovolsky	54	Vladimir Titov
25	Viktor Patsayev	55	Alexander Alexandrov
26	Vasily Lazarev	56	Vladimir Solovyov
27	Oleg Makarov	57	Oleg Atkov
28	Pyotr Klimuk	58	Igor Volk
29	Valentin Lebedev	59	Vladimir Vasyutin
30	Yuri Artyukhin	60	Alexander Volkov

Background

For the Vostok flights, as there was only one cosmonaut onboard, the order of flight was obvious. However, with the advent of multi-man crews, a system had to be devised to determine ranking in the event of two or more cosmonauts making a first spaceflight together, as happened with Voskhod 1, when three rookie cosmonauts flew together. A priority system was used based on the crew assignments, as follows:-

Mission	Priority
Vostok 1 -6	N/A
Voskhod 1	Commander; researcher; physician.
Voskhod 2	Commander; co-pilot.
Soyuz 1 – 11	Commander; Flight-engineer; test-engineer.
Soyuz 12 – to date	Commander; Flight-engineer; research-cosmonaut.

After making a first spaceflight, each cosmonaut receives various awards *(see section entitled 'Soviet Honours and Awards' for more details.)* These include the Pilot-Cosmonaut of the USSR medal, which is presented with a book and a Pilot-Cosmonaut number. Thereafter, the cosmonaut will often be referred to as, for example, 'Cosmonaut number 7, Vladimir Komarov . . .'

1.03: Intercosmos cosmonauts in order of first spaceflight

The following table shows each Intercosmos cosmonaut in order of first spaceflight. For full details of the reasons for the flight order, see section entitled 'The Intercosmos programme.'

1	Vladimir Remek	Czechoslovakia
2	Miroslaw Hermaszewski	Poland
3	Sigmund Jähn	German Democratic Republic
4	Georgi Ivanov	Bulgaria
5	Bertalan Farkas	Hungary
6	Pham Tuan	Vietnam
7	Arnaldo Tamayo-Méndez	Cuba
8	Jugderdemidiin Gurragcha	Mongolia
9	Dumitru Prunariu	Romania
10	Jean-Loup Chrétien	France
11	Rakesh Sharma	India

1.04: Crew Assignments

Introduction

The Soviets never reveal the names of the back-up crew for a mission until the cosmonauts involved make their own first flight. The only exceptions have been for the more open flights i.e. the ASTP and Intercosmos joint missions. Unfortunately, it is often the case that if a veteran cosmonaut makes a further flight, back-up duties carried out between the previous and current flight are not revealed.

This policy has meant that there are gaps in the known back-up assignments, and this is further complicated by the occasions where crews have been switched e.g. Soyuz 11, or missions cancelled e.g. Vostok 7, Soyuz 12.

Support crews appear to have existed for all missions except those carried out under the Intercosmos programme, but their existence has only recently been officially confirmed, and very few assignments have so far been identified.

The following table presents an analysis of the **known** prime, back-up and support crews, using information drawn from official Soviet sources. Those shown in *italics* are those which are strongly suspected, but for which official confirmation is awaited *(see the relevant biographies for more details.)*

Certain aspects of crewing policy are examined in the sections entitled 'Voskhod 3 Crewing'; 'Soyuz 1/2 Crewing'; 'Salyut 1 Crewing'; 'Salyut 2/Cosmos 557 Crewing'; 'Salyut 3 Crewing'; 'Salyut 4 Crewing'; 'Salyut 5 Crewing'; 'Salyut 6 Crewing'; 'Salyut 7 Crewing'; 'ASTP Crewing'; and 'Intercosmos Crewing'.

Crewing Table

Mission	Prime	Back-up	Support
Vostok 1	Gagarin	Titov, G	Grigori
Vostok 2	Titov, G	Nikolayev	*Komarov*
Vostok 3	Nikolayev	Bykovsky	Grigori / Shonin
Vostok 4	Popovich	Komarov	Volynov
Vostok 5	Bykovsky	Volynov	Komarov
Vostok 6	Tereshkova	*Tortchillova*	??
Vostok 7 (cancelled)	*Yegorov*	??	??
Voskhod 1	Komarov Feoktistov Yegorov	Volynov Makarov Lazarev	Shatalov ?? ??
Voskhod 2	Belyayev Leonov	Dmitri Khrunov	*Filipchenko* Gorbatko/ *Shonin*
Voskhod 3 (cancelled)	*Volynov* *Shonin*	*Shatalov* *Beregovoi*	?? ??
Soyuz 1	Komarov	Gagarin	??
Soyuz 2 (cancelled)	*Bykovsky* *Yeliseyev* *Khrunov*	*Belyayev/Nikolayev* *Kubasov* *Gorbatko*	?? ?? ??
Soyuz 2	Unmanned	Unmanned	
Soyuz 3	Beregovoi	Shatalov	Volynov

Soyuz 4	Shatalov	Filipchenko	??
Soyuz 5	Volynov Yeliseyev Khrunov	Shonin Kubasov Gorbatko	?? ?? ??
Soyuz 6	Shonin Kubasov	Nikolayev Sevastyanov	?? ??
Soyuz 7	Filipchenko Volkov, V Gorbatko	Nikolayev Sevastyanov Grechko	?? ?? ??
Soyuz 8	Shatalov Yeliseyev	Nikolayev Sevastyanov	?? ??
Soyuz 9	Nikolayev Sevastyanov	Lazarev Grechko	Filipchenko *Kubasov*
Soyuz 10	Shatalov Yeliseyev Rukavishnikov	*Dobrovolsky* Volkov, V *Patsayev*	?? ?? ??
Soyuz 11	Dobrovolsky Volkov, V Patsayev	*Shatalov* *Yeliseyev* ??[1]	?? ?? ??
Soyuz 12 (cancelled)	Leonov Kubasov	*Filipchenko* *Rukavishnikov*	?? ??
Soyuz 12	Lazarev Makarov	Gubarev Grechko	?? ??
Soyuz 13	Klimuk Lebedev	*Bykovsky*[2] ??[3]	?? ??
Soyuz 14	Popovich Artyukhin	Volynov *Zholobov*	*Zudov* *Rozhdestvensky*
Soyuz 15	Sarafanov Demin	*Volynov* *Zholobov*	*Zudov* *Rozhdestvensky*
Soyuz 16	Filipchenko Rukavishnikov	Romanenko Ivanchenkov	?? ??
Soyuz 17	Gubarev Grechko	*Klimuk* *Sevastyanov*	*Kovalyonok* ??
April 5 anomaly	Lazarev Makarov	Klimuk Sevastyanov	*Kovalyonok* ??
Soyuz 18	Klimuk Sevastyanov	Kovalyonok ??	?? ??

Soyuz 19 ASTP	Leonov Kubasov	Dzhanibekov Andreyev	none none
Reserve ASTP Soyuz (not flown)	Filipchenko Rukavishnikov	Romanenko Ivanchenkov	none none
Soyuz 20	Unmanned	Unmanned	—
Soyuz 21	Volynov Zholobov	Zudov Rozhdestvensky	*Gorbatko* *Glazkov*
Soyuz 22	Bykovsky Aksyonov	Malyshev Strekalov	Popov ??
Soyuz 23	Zudov Rozhdestvensky	Gorbatko Glazkov	*Berezovoi* ??
Soyuz 24	Gorbatko Glazkov	Berezovoi ??	?? ??
Soyuz 25	Kovalyonok Ryumin	Romanenko *Ivanchenkov*	*Lyakhov* *Lebedev*
Soyuz 26	Romanenko Grechko	Kovalyonok Ivanchenkov	*Lyakhov* *Ryumin*
Soyuz 27	Dzhanibekov Makarov	Kovalyonok Ivanchenkov	?? ??
Soyuz 28	Gubarev Remek	Rukavishnikov Pelčak	none none
Soyuz 29	Kovalyonok Ivanchenkov	Lyakhov Ryumin	*Popov* *Lebedev*
Soyuz 30	Klimuk Hermaszewski	Kubasov Jankowski	none none
Soyuz 31	Bykovsky Jähn	Gorbatko Köllner	none none
Soyuz 32	Lyakhov Ryumin	Popov Lebedev	*Illarionov*[4] *Strekalov*
Soyuz 33	Rukavishnikov Ivanov	Romanenko Alexandrov	none none
Soyuz 34	Unmanned	Unmanned	Unmanned
Soyuz T-1	Unmanned	Unmanned	Unmanned
Soyuz 35	Popov Ryumin	Titov, V Strekalov	*Kovalyonok* *Savinykh*

Soyuz 36	Kubasov	Dzhanibekov	none
	Farkas	Magyari	none
Soyuz T-2	Malyshev	Kizim	??
	Aksyonov	Makarov	??
Soyuz 37	Gorbatko	Bykovsky	none
	Tuan	Liem	none
Soyuz 38	Romanenko	Khrunov	none
	Tamayo-Méndez	Lopez-Falcon	none
Soyuz T-3	Kizim	Lazarev	??
	Makarov	Ivanchenkov	??
	Strekalov	Savinykh	??
Soyuz T-4	Kovalyonok	Titov, V	??
	Savinykh	Strekalov	??
Soyuz 39	Dzhanibekov	Lyakhov	none
	Gurragcha	Ganzorig	none
Soyuz 40	Popov	Romanenko	none
	Prunariu	Dediu	none
Soyuz T-5	Berezovoi	Titov, V	*Lyakhov*
	Lebedev	Strekalov	*Alexandrov*
Soyuz T-6	Dzhanibekov	Kizim	none
	Ivanchenkov	Solovyov	none
	Chrétien	Baudry	none
Soyuz T-7	Popov	Vasyutin	??
	Serebrov	Savinykh	??
	Savitskaya	*Irina*	??
Soyuz T-8	Titov, V	*Lyakhov*	*Kizim*
	Strekalov	*Alexandrov*	*Solovyov*
	Serebrov	*Savinykh* [5]	??
Soyuz T-9	Lyakhov	*Kizim*	*Vasyutin*
	Alexandrov	*Solovyov*	*Savinykh*
Launch Abort	Titov, V	Kizim	*Vasyutin*
	Strekalov	Solovyov	*Savinykh*
Soyuz T-10	Kizim	Vasyutin	*Dzhanibekov*
	Solovyov	Savinykh	*Volk*
	Atkov	??	??
Soyuz T-11	Malyshev	Berezovoi	none
	Strekalov	Grechko	none
	Sharma	Malhotra	none

Soyuz T-12	Dzhanibekov	Vasyutin	??
	Savitskaya	*Irina*	??
	Volk	Savinykh	??
Soyuz T-13 (cancelled)	Vasyutin	*Titov* [6]	??
	Savinykh	*Strekalov* [6]	??
	Volkov, A	*Serebrov* [6]	??
Soyuz T-13	Dzhanibekov	*Vasyutin*	??
	Savinykh	*Grechko*	??
Soyuz T-14	Vasyutin	*Titov* [6]	??
	Grechko	*Strekalov* [6]	??
	Volkov, A	*Serebrov* [6]	??

Notes

1 Unnamed cosmonaut selected in 1963 *(see section entitled 'Salyut 1 Crewing' for more details.)*

2 This is highly speculative – see biography.

3 See Lebedev's biography.

4 This is highly speculative – see biography.

5 Possibly confirmed – see biography.

6 This is highly speculative – see biography.

1.05: Voskhod 3 Crewing

Introduction

In his autobiography entitled 'Trudniye Dorogi Kosmosa' published in 1978, Vladimir Shatalov revealed that a *third* mission in the Voskhod series was planned for the early summer of 1966, involving a two-man team of cosmonauts in a long-duration mission. It has been suggested that the flight would have lasted 7 or 8 days, and was intended as a replacement for Boris Yegorov's cancelled Vostok 7 mission.

Confirmation of the Voskhod 3 mission came in Yuri Gagarin's diary, published in 1983.

Background

Voskhod 3 appears to have been cancelled at the very last moment. According to Gagarin's diary, it was a bureaucratic decision which, not suprisingly, was unpopular with the cosmonaut team. Shatalov's explanation was that the cancellation was ordered because the new Soyuz programme was nearly ready.

In a Soviet article published in 1981, it was revealed that the Voskhod programme originally included five missions:

Mission	Date	Objective
Voskhod 1	1964	multi-manned flight.
Voskhod 2	1965	first spacewalk.
Voskhod 3	1966	long-duration, 12 day flight.
Voskhod 4/5	1967	rendezvous between 2 Voskhods, using Polyot engines. In preparation for this mission, two experimental Polyot satellites were launched in 1963 and 1964 to test the new engines which the Voskhods would use to change orbit.

The same source reported that Voskhod 3 was cancelled by Korolyov's successor, Vasili Mishin, and the mission objectives were incorporated in the 17 day long Soyuz 9 flight in 1970.

Following the cancellation of Voskhod 3, there were no further acknowledged Voskhod flights, and the next Soviet manned mission was the first flight of the new Soyuz craft in April 1967. The ambitious US Gemini programme had been very successful and so the pressure to achieve a rendezvous and possible docking with Voskhod spacecraft was removed. This, and the fact that the more advanced Soyuz spacecraft was nearly ready for first flight, was the most likely cause of the cancellation of further Voskhod missions. In addition, there was a new political scene in the Soviet Union following the overthrow of Kruschev, and the pressure for Soviet space spectaculars is thought to have abated.

It appears however, that a further Voskhod spacecraft may have been launched, as Cosmos 110 on 22 February 1966. According to TASS, the craft carried two dogs, Veterok and Ugolek, and the aim of the flight was to study the longer term effects of space radiation and weightlessness. The spacecraft flew in an orbit with an apogee of over 900km, inside the inner radiation belt. The two dogs returned to Earth after a flight lasting 21.75 days.

This may well have been a test-flight for the planned Voskhod 3 mission, carried out just prior to the cancellation of the Voskhod programme.

Crewing

A training group was formed, probably around April 1965, with the mission cancelled in early 1966.

The four cosmonauts named as being part of this group were Beregovoi, Shatalov, Volynov and Shonin, and pictures of both teams training in Voskhod-style capsules have been released. It is thought that the prime crew would have been Volynov and Shonin, with Beregovoi and Shatalov as the back-ups.

(Volynov was the most likely prime commander, as he had already served as back-up for the Vostok 5 and Voskhod 1 missions.)

Mission	Prime	Back-up	Support
Voskhod 1	Komarov	Volynov	Shatalov
	Feoktistov	Makarov	?
	Yegorov	Lazarev	?
Voskhod 2	Belyayev	Dmitri	*Filipchenko*
	Leonov	Khrunov	Gorbatko/
			Shonin
Voskhod 3	*Volynov*	*Shatalov*	?
	Shonin	*Beregovoi*	?
Voskhod 4	?	?	?
	?	?	?
Voskhod 5	?	?	?
	?	?	?

As the Voskhod 4/5 mission never left the drawing-board, it seems unlikely that any cosmonauts actually trained for it.

Conclusion

It is clear from the official Soviet evidence that a Voskhod 3 mission was planned, and that crews were assigned and trained for the mission, only for it to be cancelled at the last moment. It would have been a long-duration flight, lasting anywhere between 7 – 22 days, and would have been followed by a rendezvous between two Voskhod spacecraft.

The Voskhod programme was cancelled before completion, the crews re-assigned, and the mission objectives incorporated in the new Soyuz programme.

Postscript

A further mystery concerning the Voskhod programme arose following publication in 1982 of Konstantin Feoktistov's book 'About Spaceships.' The book seems to convey that Komarov and Volynov were both chosen to command the Voskhod 1 mission, and they were then joined by six scientists. The original selections seem to have been made in two groups of four, neither of which was designated prime or back-up:

Komarov	Volynov
Feoktistov	Engineer ?
Engineer ?	Yegorov
Lazarev	Doctor

Why there were four men in each group, and why the crews changed was not explained, although apparently the prime doctor withdrew at some stage. Naturally, the identities of the doctor and the engineers were not disclosed, although it seems certain that Oleg Makarov was one of the engineers.

One possibility of course, is that the mission planners simply changed the composition of the crews to enable them to work with both commanders and to find the best possible permutation. But why were there four men in each group ?

1.06: Soyuz 1/2 Crewing

Introduction

At the time of the Soyuz 1 launch, Moscow was alight with rumours about a forthcoming space spectacular. The rumours were very specific in nature, and predicted the launch dates, that two new Soviet spacecraft would be launched within 48 hours of each other, and that they would achieve the World's first manned docking and exchange of crews. Crewmembers would included Belyayev, Feoktistov and Bykovsky.

As expected, a new spacecraft **was** launched on 23 April 1967, commanded by Vladimir Komarov. Confirmation that it was the first of a double launch appeared to come with the name of the spacecraft: Soyuz 1. Soyuz meant 'Union', which implied a docking, and the fact that it was called Soyuz **1** indicated that a Soyuz 2 was already in existence.

However, the second launch never came. Komarov returned to Earth after just over a day in orbit, and was killed when the spacecraft's parachute tangled following re-entry.

It therefore appeared as if Soyuz 1 was simply a short test flight, and that a second launch had never been intended. The rumours had all been wrong.

Background

However, from 1977 onwards, some ten years after the event, things became much clearer. An official Soviet photograph released in 1977 showed Gagarin, the back-up to Komarov on Soyuz 1, posing with a group of cosmonauts who were training for a mission. They were – Khrunov, Komarov, Yeliseyev and Bykovsky. Khrunov and Yeliseyev wore spacesuits, whilst Komarov and Bykovsky wore flight overalls.

As Komarov was in the photograph, it had to have been taken prior to April 1967. As the four men were obviously training together, the implication was clear. These were the two crews of the intended Soyuz 1/2 mission. Yeliseyev and Khrunov were to have spacewalked from Soyuz 2, commanded by Bykovsky, to Soyuz 1, commanded by Komarov.

Two years later, the intended mission was carried out by the Soyuz 4/5 spacecraft – only the commanders had changed. Komarov was dead, and Bykovsky was no longer available, as in mid-1967, he was assigned to the Zond training group, and began preparing for a flight to the Moon.

In 1981, an official Soviet source mentioned in passing that Kubasov and Yeliseyev were among those cosmonauts selected in September 1966, 'in view of the forthcoming first two flights of Soyuz, Soyuz **2** having a crew of a pilot, an onboard engineer and a research engineer.' This then, was the long-awaited official confirmation of the Soyuz 2 flight. The report went on to say that the orbital rendezvous planned for Soyuz 1 and 2 in April 1967 wasn't realised until January 1969 i.e. Soyuz 4 and 5.

In 1984, Kubasov revealed that he had trained with Gorbatko and Gagarin in the period August 1966 – April 1967 for a flight onboard 'the new Soyuz spaceship, which was a promising but "untamed" vehicle . . .' In fact, he was assigned as back-up to Yeliseyev almost immediately after joining the cosmonaut team.

Looking back to news items released following the launch of Soyuz 1, Western reports stated at the time that Bykovsky was ready and waiting to command the second spacecraft. Other reports said that Belyayev and Bykovsky were among five cosmonauts ready to lift-off. In addition, a report was received in London **before** the launch of Soyuz 1, accurately giving the launch date and Komarov's name. The same correspondent predicted a second Soyuz, carrying Belyayev and Bykovsky.

It has therefore been speculated that Belyayev was the back-up commander for Soyuz 2, but no firm evidence exists. However, two pre-launch photographs have been released showing Komarov and the prime Soyuz 2 crew together with Andrian Nikolayev, leading some observers to suggest that Nikolayev was the back-up commander, and not Belyayev.

However, if Nikolayev was the back-up, he should have been re-assigned together with Kubasov and Gorbatko to form the Soyuz 5 back-up crew. In fact, he was actually back-up commander for the Soyuz 6, 7 and 8 missions, and his absence from the Soyuz 5 back-up crew would seem to indicate that he was NOT the Soyuz 2 back-up commander.

The circumstantial evidence for Belyayev is, however, slightly stronger. He was put in charge of the induction of the 1967 group of cosmonauts in May 1967 i.e. one month after the Soyuz 2 mission was cancelled. In addition, he is believed to have started training for a Zond circumlunar flight in late 1967. The flight was probably scheduled for late 1968, and therefore his position in the Soyuz 5 back-up crew would have had to have been given to someone else.

Crewing

The following table shows the flight crew assignments for the Soyuz 1/2 mission, the Soyuz 2/3 test-flight, and the Soyuz 4/5 mission which eventually flew the mission originally intended for Soyuz 1/2:

Mission	Prime	Back-up
Soyuz 1	Komarov	Gagarin
Soyuz 2 (cancelled)	*Bykovsky* *Yeliseyev* *Khrunov*	*Belyayev/Nikolayev* *Kubasov* *Gorbatko*
Soyuz 2	Unmanned	Unmanned
Soyuz 3	Beregovoi	Shatalov (also see Volynov's biography)
Soyuz 4	Shatalov	Filipchenko
Soyuz 5	Volynov Yeliseyev Khrunov	Shonin Kubasov Gorbatko

Conclusion

With the benefit of hindsight, it is now clear, despite official Soviet denials, that a joint Soyuz 1/2 mission was always intended. Furthermore, following a docking, an exchange of crews would have taken place. The mission profile was eventually flown by the crews of

Soyuz 4/5 in January 1969, following a test flight involving Beregovoi onboard Soyuz 3, during which he attempted (unsuccessfully) to dock with the unmanned Soyuz 2.

The only question remaining is that of what actually went wrong onboard Soyuz 1? An onboard failure of some kind must have occurred very early in the flight – early enough to cancel the planned Soyuz 2 launch. Western analysts have suggested that the spacecraft was tumbling wildly, or that at best, there was a serious problem with the Soyuz 1 guidance control systems. Again, the Soviets have always denied that anything went wrong on the mission until the parachute was deployed.

Therefore, most of the mystery concerning the Soyuz 2 mission appears to have been resolved, whereas that surrounding Soyuz 1 remains.

Postscript

Another mystery concerns training pictures released after Komarov's death, which showed him training in the same type of spacesuit as Khrunov and Yeliseyev. This would indicate that he had trained for an EVA of some description in connection with Soyuz 1/2. But there is no apparent reason for him to have conducted such an operation. During the crew transfer, he would have sealed off the descent module, while Khrunov and Yeliseyev entered the orbital module, and he would therefore not have needed a spacesuit. And on the duplicate Soyuz 4/5 mission, two years later, Vladimir Shatalov did not have a spacesuit.

Therefore, it raises the question – did Soyuz 1 only have a descent module i.e. no orbital module? If that was the case, then when Yeliseyev and Khrunov transferred to Soyuz 1 by external EVA, Komarov would have had to depressurize the descent module, hence the need for his spacesuit. Because of the possible absence of an orbital module, it has even been suggested that Soyuz 1 may in fact have been a Zond, and it is noteworthy that no pictures of Soyuz 1 have ever been released.

1.07: Salyut 1 crewing

Introduction

Salyut 1 was the world's first manned space-station, launched on 19 April 1971 to mark the 10th anniversary of Yuri Gagarin's flight onboard Vostok 1 on 12 April 1961. (Salyut is the Russian word for Salute.) The crewing of Salyut 1 has become one of the most complex and intriguing aspects of the Soviet space programme. With the exception of Volkov on Soyuz 10, the back-up crews for the Salyut 1 missions have still not been revealed. In addition, it appears that one of the crews scheduled to visit Salyut 1 was stood down due to illness, and this, combined with the Soyuz 10 failure, seriously disrupted the flight schedule.

This section details the first two missions to Salyut 1, and examines the crewing.

Soyuz 10

The Soyuz 10 mission, launched on 23 April 1971, has always been described by the Soviets as a wholly successful flight, but it is clear that this was not the case.

Following the first ever docking with a space-station, the crew failed to enter the Salyut. The Soviets have stated that it was never their intention for the station to be manned during this initial step, but in reality it would seem likely that there was actually some problem with the transfer hatch, preventing the crew from entering the Salyut.

The mission had to be aborted, and the crew returned to Earth at the very first opportunity to do so and still return in the usual recovery area, even though it meant landing in the dark. This was the first time the Soviets had made a night landing, and this appeared to confirm that there was either trouble in Soyuz 10, or such dependence on Salyut systems that when these were unavailable, there was no point in prolonging the mission. It is also noteworthy that if all previous Soyuz flights are plotted on a graph to compare the hour of launch with the number of days in flight until recovery, a very linear relationship is found. On the basis of this type of estimate, the pre-dawn launch of Soyuz 10 suggested a 30-day flight. However, it remained in orbit for just 2 days.

Soyuz 11

Soyuz 11 was launched on 6 June 1971, just six weeks after the Soyuz 10 failure. This time, the crew were successful in boarding the Salyut, and became the first men to work in a manned space-station. They spent 23 days onboard, carrying out a variety of experiments before returning to Earth. Tragically, a pressure equalization valve opened prematurely while the Soyuz descent module was still above the atmosphere, and the crew were killed following the loss of cabin oxygen and pressure.

Any further missions to the Salyut were immediately cancelled.

It should be noted that an analysis of the launch and landing window data indicates that the Soyuz 11 mission was also cut short.

Crewing – Background

During 1969, a new training group was formed for the Salyut space-station programme. The team was a blend of experienced cosmonauts and rookies, and only one new cosmonaut was needed. Viktor Patsayev was recruited specially for the programme in 1969, and joined the Salyut training group immediately after selection.

In the Soviet film 'Steep Road to Space', the training group is shown standing around a globe of the Earth. Nine cosmonauts appear, and it seems certain that they are the three Salyut 1 crews. Eight men can be readily identified: Dobrovolsky, Kubasov, Leonov, Patsayev, Rukavishnikov, Shatalov, Volkov, and Yeliseyev. A ninth man has not yet been positively identified, although it is believed that he was a military-engineer, as he was pictured wearing a military uniform, with no Air Force pilot's wings.

The following are the only officially confirmed Salyut 1 assignments:

Mission	Prime	Back-up
Soyuz 10	Shatalov	
	Yeliseyev	Volkov
	Rukavishnikov	
Soyuz 11	Dobrovolsky	
	Volkov	
	Patsayev	see note

Note:
According to one Soviet source, the back-up to Patsayev was a cosmonaut selected in 1963, who was still in training in 1981 for a first spaceflight.

In addition Soviet reports have spoken of a third crew:

Leonov
Kubasov
an unnamed military engineer

Leonov and Kubasov's involvement with Salyut 1 has been officially confirmed, and they have both been pictured undergoing Salyut training. In July 1975, at the time of ASTP, rumours circulated in Moscow that Leonov had actually been assigned as the commander of Soyuz 11 but had been forced to step down when Kubasov fell ill with a communicable disease a few weeks prior to the mission. The back-up crew therefore flew the mission instead.

In 1979, apparent confirmation came in a Soviet book about ASTP which reported: 'After the Soyuz 6 mission, Valery Kubasov worked in the design bureau and prepared for another orbital flight with a more complicated assignment. But just before the flight, the doctors discovered he had a lung condition and disqualified him.'

'The months that dragged out agonisingly after they struck me off the list of cosmonauts were the hardest ones in my life,' he later recalled. 'I felt infinitely sorry about all the work put in by the big team of specialists who had trained me for the flight...They had plans and goals of their own, and while I had not disrupted them, I had undoubtedly postponed them.' He made a full recovery, and was eventually placed back on the cosmonaut team.

It is clear that Leonov and Kubasov were intended to fly to Salyut 1, but the identity of the third cosmonaut in their crew has been a matter of much speculation. Some reports stated that he was a military engineer, and this led to speculation that it was Khrunov or Volynov. However, in 1975, a photograph appeared which caused much controversy. It showed three men training in a Soyuz simulator and was therefore taken prior to the Soyuz 11 disaster. Leonov and Kubasov are clearly shown, and part of the face (little more than the forehead) of a third man is also shown.

It is widely believed that this picture depicts the Leonov Salyut 1 crew. Many attempts have been made by Western observers to identify the third man, and it was originally thought that he was Yuri Artyukhin. However, it was later learnt that Artyukhin was a CapCom onboard a tracking ship during the Salyut 1 mission, and could not therefore have been a crewmember.

It was then suggested that the third cosmonaut was Viktor Patsayev, and later, that it was Nikolai Rukavishnikov. The latter appeared to be confirmed several years later, when another photograph was released clearly showing Rukavishnikov, Leonov and Kubasov training together.

Crewing

It would appear that only two flights to Salyut 1 were originally planned, and Dobrovolsky, Volkov and the unknown military cosmonaut were given the back-up assignments for both missions.

Whilst it is clear that Leonov, Kubasov and another cosmonaut should have flown to Salyut 1, it is uncertain as to whether they should have been the first or second crew.

This section therefore examines both alternatives:-

First crew

When Kubasov became ill, he, Leonov and Rukavishnikov were stood down. The Shatalov crew, intended to fly second, were given the primary mission. However, because of Rukavishnikov's Salyut expertise, it was essential to fly him on the first mission to board Salyut 1. He was therefore moved into the replacement Soyuz 10 crew, bumping Patsayev, who in turn, bumped the unknown cosmonaut from Dobrovolsky's crew. The Dobrovolsky crew were also moved up, and were given the Soyuz 11 assignment. Shatalov and Yeliseyev may have been assigned as the back-ups, as they were the only other Salyut-trained cosmonauts available.

This theory would seem to be supported by later Soviet accounts which suggested that Rukavishnikov was 'a late addition to the Soyuz 10 crew' and that Patsayev was 'a last-minute addition to the Dobrovolsky crew in 1971.'

Following Kubasov's recovery, he and Leonov were teamed with the engineer bumped from Dobrovolsky's crew for a third mission following the Soyuz 10 failure. One Soviet source has confirmed that a second occupation of Salyut 1 was cancelled, and that the crew of Leonov, Kubasov and a military engineer were to have stayed in the station for a month-long flight. It is believed that the mission would have been launched in mid-August 1971.

Original plan:

Mission	**Prime**	**Back-up**
Soyuz 10	Leonov Kubasov Rukavishnikov	Dobrovolsky Volkov unknown
Soyuz 11	Shatalov Yeliseyev Patsayev	Dobrovolsky Volkov unknown

following Kubasov's illness:

Soyuz 10	Shatalov Yeliseyev Rukavishnikov	Dobrovolsky Volkov Patsayev
Soyuz 11	Dobrovolsky Volkov Patsayev	*Shatalov* *Yeliseyev* unknown 1963 cosmonaut
Soyuz 12	Leonov Kubasov unknown	?? ?? ??

Second crew

Following the Soyuz 10 failure, the Soyuz 11 crew of Leonov, Kubasov and Patsayev should have flown. But Kubasov became ill, and the crew were stood down. They were replaced by the Dobrovolsky crew, but Patsayev, specially chosen for the Salyut 1 programme, was switched to the Dobrovolsky crew, thus bumping the military engineer. Shatalov and Yeliseyev may have been assigned as the back-ups, as they were the only other Salyut-trained cosmonauts available.

This theory would confirm earlier reports that Soyuz 11 was flown by its back-up crew, and would explain the photograph apparently showing Leonov, Kubasov and Patsayev training together. In addition, Patsayev was said to be 'a last-minute addition to the Dobrovolsky crew in 1971' and this would be explained by a switch from the Leonov crew.

Following Kubasov's recovery, he and Leonov were teamed with the engineer bumped from Dobrovolsky's crew for a third mission – see above.

Original plan:

Mission	**Prime**	**Back-up**
Soyuz 10	Shatalov Yeliseyev Rukavishnikov	Dobrovolsky Volkov unknown
Soyuz 11	Leonov Kubasov Patsayev	Dobrovolsky Volkov unknown

following Kubasov's illness:

Soyuz 11	Dobrovolsky Volkov Patsayev	*Shatalov* *Yeliseyev* unknown 1963 cosmonaut
Soyuz 12	Leonov Kubasov unknown	?? ?? ??

Conclusion

Even now, so long after the event, it cannot be said with any certainty whether either crewing theory is correct. However, when each piece of evidence is examined, it would appear that the case for Leonov and Kubasov being part of the first crew is marginally stronger than the case for them being on the original Soyuz 11 crew.

1.08: Salyut 2 and Cosmos 557 Crewing

Introduction

It is believed that there were three attempts at putting a second successful Salyut into operation in the period July 1972 – May 1973, all of which failed. This section examines each of those attempts, and analyses the crewing together with the possible reasons for failure.

1972 Salyut attempt

It is believed that an unsuccessful attempt to launch a Salyut was made on 30 July 1972. In March of that year, a strange report was released by *Agence France Press*. Quoting 'generally well-informed circles' in Moscow, the news agency reported that the Soviet Union would launch a new Salyut the following May, and that this would stay in orbit for two months.

Two two-man Soyuz crews would man the space-station on a relay basis, each staying onboard for 30 days. The crews would consist of two military officers – two rookies known only as Vinogradov and Voronov – and two civilians – Vitaly Sevastyanov and Vasily (sic) Kubasov. The report also identified Pavel Popovich as a back-up crew member.

The report is a strange one, because many of its elements were correct. It said that the Soyuz had been redesigned to carry only two men, and that they would be required to wear spacesuits during the launch and return phases of the mission. This was not officially known at the time, as no flights had been made since Soyuz 11. Of the men identified, Kubasov and Sevastyanov were definitely involved in the civilian Salyut programme, whilst Popovich was involved in the military Salyut programme. Finally, Voronov and Vinogradov are real cosmonauts – the American ASTP astronauts actually met them at Star Town.

Therefore it would seem that the report was probably accurate, and that two crews were to have been launched to a Salyut in 1972. Originally scheduled for May, the Salyut launch was apparently delayed until July. Many questions remain unanswered however, and it is puzzling that Vinogradov and Voronov have still not flown.

Salyut 2

Salyut 2 was launched on 3 April 1973, and for reasons not yet explained by the Soviets was never manned. Its mission was concluded after only three weeks, and it re-entered the Earth's atmosphere a month later.

The space-station was launched into a 162 × 134 mile (260 × 215 km) orbit inclined at 51.6 degrees and with a period of 89 minutes. Various reports said that a Soyuz launch would soon follow, and on two occasions, the space-station was in a position for rendezvous, but no launch occurred. On 8 April, Salyut 2's orbit was raised to 167 × 154 miles (268 × 248 km), above an appropriate rendezvous orbit, and Western observers concluded that whatever had delayed the Soyuz launch was more serious than originally thought. It was suggested by some that solar flare activity on 4 and 5 April had prevented the launch rather than equipment failure, but when 11 April came and there was still no launch, it was concluded that either the Salyut or the Soyuz was having major difficulty.

Some Western observers believe that Salyut 2 suffered severe guidance and control problems, and it was noted that the space-station's initial orbit was higher and more eliptical than Salyut 1's, possibly due to poor performance of the Proton D booster. Numerous fragments detected in the orbital path suggested that the Proton had exploded, although in retrospect, it now seems likely that these early pieces of debris were no more than the routine releasing of equipment and window covers.

On 14 April, Salyut 2 was reported to have undergone a 'catastrophic malfunction' which not only ripped off the solar panels but also the rendezvous radar and radio-transponder, leaving the space-station tumbling without telemetry return. The Salyut may have separated into many pieces, some large enough to be tracked, but most were rather small and decayed quickly. The Soviets have never explained what happened to the space-station, but it is believed that either an explosion or a misfiring thruster was to blame. The most widely held theory is that the Proton upper stage exploded, and its debris damaged the space-station.

On 28 April, TASS reported that Salyut 2 'had concluded the programme of flight,' and although the official statement said it had completed its mission, the word 'successfully' was omitted, giving an indication that it was officially viewed as a failure. The main body of the station decayed through airdrag on 28 May 1973, and re-entered near Australia.

Salyut 2 Crewing

The day after Salyut 2 was launched, cosmonaut Yevgeny Khrunov announced that cosmonauts were engaged in preparations for new flights, presumably to link up with Salyut 2, and on 6 April Victor Louis of the London Evening News reported that a Soyuz spacecraft was ready for launch.

On 10 April 1973, during a lecture at the Helsinki University, Academician Boris Petrov discussed Salyut 2 and said that as it was more automatic than Salyut 1, only 2 men would be needed onboard, instead of the 3 on the previous station.

It appears that Salyut 2 was the first attempt at putting a military Salyut into orbit *(see section entitled 'Military and Civilian Salyut programmes' for more details)* and that its intended mission was later carried out by Salyut 3. Three crews are known to have trained for Salyut 3, and some photographs indicate that they might have been in training for Salyut 2 as well. The three Salyut 3 crews were:

Pavel Popovich and Yuri Artyukhin;
Boris Volynov and Vitaly Zholobov;
Gennady Sarafanov and Lev Demin.

No firm evidence of these crews being assigned to Salyut 2 is yet available.

Cosmos 557

Cosmos 557 was launched on 11 May 1973, into a 140 × 135 mile (226 × 218 km) orbit inclined at 51.6 degrees and with a period of 89.1 minutes. The launch came shortly after the failure of Salyut 2 – in fact, less than six weeks after the Salyut 2 launch – and has been identified in the West as a successfully orbited Salyut. Its telemetry resembled Salyut 1, rather than Salyut 2, and it is therefore thought that Cosmos 557 was a civilian Salyut *(see section entitled 'Military and Civilian Salyut programmes' for more details.)* However, it appears that the space-station suffered a propulsion or command-sequence failure which occurred so soon after orbital insertion that it was given the Cosmos 557 designation. The failure rendered it useless for human occupation, and it rapidly decayed in orbital altitude, re-entering the Earth's atmosphere on 22 May 1973.

Tracking ships which had been deployed for the expected manned flights to Salyut 2 were reported as heading back to their home ports before Cosmos 557 decayed. It is interesting to speculate whether it was apparent early on that no manned missions would be sent to the station, or whether there was never any intention of sending men to it. However, it seems extremely unlikely that the mission was intended to be an unmanned test of the Salyut, as the cost could not be justified. Therefore, if Cosmos 557 was a Salyut, it would seem certain that manned flights to it were planned, but they were cancelled when the station failed.

The failure of Cosmos 557 set the Salyut programme back once more, and whilst its problems were being tackled, the Soviets used the time to test-fly the new Soyuz ferry which no longer carried solar panels. An unmanned version was flown as Cosmos 573, and the first manned test was the Soyuz 12 mission. It is thought by some that these two spacecraft were originally intended for use with Cosmos 557.

Cosmos 557 Crewing

It is believed that Alexei Leonov and Valeri Kubasov were the prime crew for a mission to the Cosmos 557 Salyut. They had originally been scheduled to fly to Salyut 1, but were prevented from doing so by either Kubasov's illness or the Soyuz 11 tragedy *(see section entitled 'Salyut 1 Crewing' for more details.)* According to one Western report, shortly after the Cosmos 557 failure the Soyuz 12 launch was cancelled, and Leonov was named as commander of the Soviet spacecraft for the ASTP mission. The Soviet-American flight was scheduled for 15 July 1975, and upon being given the assignment, Leonov is reported to have complained 'But I don't even speak English !' Shatalov is then said to have replied 'That's alright – you have two years and two months to learn.' If correct, this would confirm that Leonov and Kubasov were to have flown a May 1973 Salyut mission.

In an interview in 1975, Leonov indirectly confirmed that he was in training to command a Salyut station **after** Salyut 1. 'Just to learn to operate the solar telescope I spent three months at the Crimea Observatory,' Leonov said. However, there was no solar telescope on Salyut 1. The first successful instrument was carried on Salyut 4, as witnessed by its designation: OST-1. This would appear to be further confirmation that Leonov was intended to fly a 1973 Salyut mission.

As Filipchenko and Rukavishnikov were also assigned to ASTP, it seems likely that they too had been in training with Leonov and Kubasov for a 1973 Salyut flight.

It is also possible to speculate that the other two crews were Vasily Lazarev and Oleg Makarov, and Alexei Gubarev and Georgi Grechko. Lazarev and Makarov flew the first manned Soyuz launched after the Cosmos 557 failure, and Gubarev and Grechko were the back-ups on the mission, Soyuz 12. In addition, the first crew to be launched to the next civilian Salyut were Gubarev and Grechko onboard Soyuz 17, closely followed by Lazarev and Makarov.

It therefore seems reasonable to speculate the following Cosmos 557 crewing assignments:

	Prime	Back-up	Support
1:	Leonov Kubasov	Filipchenko Rukavishnikov	Gubarev Grechko
2:	Lazarev Makarov	Filipchenko Rukavishnikov	Gubarev Grechko

Following the failure of Cosmos 557, the crews were re-assigned. The Leonov/Kubasov and Filipchenko/Rukavishnikov crews were switched to ASTP, and the Lazarev/Makarov and Gubarev/Grechko crews were given a 'fill-in' mission, Soyuz 12, which was a test of the new Soyuz ferry.

1.09: Salyut 3 Crewing

Introduction

Salyut 3 was the first successful military Salyut, launched on 25 June 1974. Only one crew managed to board the space-station, as the second mission failed to dock. This section examines the planned crewing for Salyut 3.

Crewing

It would appear that as with Salyuts 1 and 4, only two missions were planned for Salyut 3. The same crewing policy was followed, with two prime crews and one back-up crew being assigned. It is also thought that a support crew was assigned, although this has not been confirmed.

The crews were as follows:

Mission	Prime	Back-up	Support
Soyuz 14	Popovich Artyukhin	Volynov *Zholobov*	*Zudov* *Rozhdestvensky*
Soyuz 15	Sarafanov Demin	*Volynov* *Zholobov*	*Zudov* *Rozhdestvensky*

It is believed that the prime and back-up crews may have trained originally for Salyut 2 *(see section entitled 'Salyut 2/Cosmos 557 Crewing' for more details.)* Following the failure of that Salyut, the cosmonauts were re-assigned to Salyut 3.

The first mission to Salyut 3, Soyuz 14, went according to plan. However, the second mission, Soyuz 15, was aborted following a failure in the docking and rendezvous equipment. Had it been successful, it is believed that the flight would have involved a month-long stay onboard the Salyut. The back-up crew did not fly a replacement mission, and Salyut 3 was abandoned.

The Soyuz 15 back-up crew has never been officially identified, but there is supporting evidence to show that it was Volynov and Zholobov. Volynov has been named as back-up commander on Soyuz 14, and both men flew together on Soyuz 21. In addition, Zholobov has been named as a back-up flight-engineer for Salyut 3. As no mission has been specified, it is assumed that he was back-up for both planned missions. Finally, when the Salyut 1 and 4 crewing policies are examined, it appears that there was only one back-up crew in each case, and it seems reasonable to assume that the same policy existed for Salyut 3.

The Zudov and Rozhdestvensky support crew assignments have not been officially confirmed, although the two men have been linked with Salyut 3

1.10: Salyut 4 Crewing

Introduction

Salyut 4 was the second successful civilian Salyut, launched on 26 December 1974. The crewing schedule for the space-station had to be altered when the second mission was aborted during the launch phase, and this section therefore examines both the planned and the revised crewing.

Crewing

It would appear that as with Salyuts 1 and 3, only two missions were planned for Salyut 4. The same crewing policy was followed, with two prime crews and one back-up crew being assigned. It is also thought that a support crew was assigned, although this has not been confirmed.

The crews were as follows:

Mission	Prime	Back-up
1	Gubarev Grechko	Klimuk Sevastyanov
2	Lazarev Makarov	Klimuk Sevastyanov

It is believed that the Gubarev/Grechko and Lazarev/Makarov crews were originally assigned to the Cosmos 557 Salyut attempt *(see section entitled 'Salyut 2/Cosmos 557 Crewing' for more details.)* Following the failure of the Salyut, the two crews were given a 'fill-in' mission, Soyuz 12, which was a test of the new Soyuz ferry.

A new group was then formed for flights to the next civilian Salyut, and the Gubarev/Grechko and Lazarev/Makarov crews were given the prime assignments for the two planned missions. Klimuk and Sevastyanov were assigned as back-up crew for both missions, and Kovalyonok and another, as yet unknown, cosmonaut were brought in as support crew.

The first mission to Salyut 4, Soyuz 17, went according to plan. However, the second mission, Soyuz 18, was aborted during the launch phase, and was re-designated the 'April 5 anomaly.' The back-up crew of Klimuk and Sevastyanov then flew the replacement Soyuz 18, and Kovalyonok and the unknown cosmonaut moved up from support crew to back-up crew on the new mission. The revised crewing was therefore as follows:

Mission	Prime	Back-up	Support
Soyuz 17	Gubarev Grechko	*Klimuk* *Sevastyanov*	*Kovalyonok* ??
April 5 anomaly	Lazarev Makarov	Klimuk Sevastyanov	*Kovalyonok* ??
Soyuz 18	Klimuk Sevastyanov	Kovalyonok ??	?? ??

The Soyuz 17 back-up crew has never been officially identified, but there is supporting evidence to show that it was Klimuk and Sevastyanov. Most importantly, they were pictured accompanying Gubarev and Grechko on the flight to Baikonur, prior to launch, and the prime crew are usually accompanied by the back-up crew. Secondly, when the Salyut 1 and 3 crewing policies are examined, it appears that there was only one back-up crew in each case, and it seems reasonable to assume that the same policy existed for Salyut 4.

The Kovalyonok support crew assignment has not been officially confirmed, and the back-up flight-engineer for Soyuz 18 has still not been identified. At one time, it was thought that it was Valery Ryumin, based on the fact that he and Kovalyonok were the first crew to fly to Salyut 6, the next civilian space-station. However, it has since been learnt that Ryumin was a CapCom onboard a tracking ship during the Soyuz 17 and 18 missions, and could not therefore have served on the support crew.

1.11: Salyut 5 Crewing

Introduction

Salyut 5 was the second successful military Salyut, launched on 22 June 1976. The crewing schedule for the station had to be altered when the second mission failed to dock. This section therefore examines both the planned and the revised crewing.

Crewing

It would appear that as with Salyuts 1, 3 and 4, only two missions were planned for Salyut 5. However, a new crewing policy was introduced, and instead of only one back-up crew for both missions, separate back-up crews were assigned. This policy allowed the back-up crew to fly as prime on the next mission. It is also thought that support crews were assigned, although this has not been confirmed.

The crews were as follows:

Mission	Prime	Back-up
1	Volynov Zholobov	Zudov Rozhdestvensky
2	Zudov Rozhdestvensky	Gorbatko Glazkov

The two prime crews had originally trained for Salyut 3 *(see section entitled 'Salyut 3 Crewing' for more details.)* Following the completion of missions to that Salyut, the cosmonauts were re-assigned to Salyut 5.

The first mission to Salyut 5, Soyuz 21, went according to plan. However, the second mission, Soyuz 23, was aborted following a failure in the docking and rendezvous equipment. (Soyuz 22 was a solo Soyuz, and was not part of the Salyut programme.) The back-up crew of Gorbatko and Glazkov then flew a replacement mission, Soyuz 24. The revised crewing was therefore as follows:

Mission	Prime	Back-up	Support
Soyuz 21	Volynov Zholobov	Zudov Rozhdestvensky	*Gorbatko* *Glazkov*
Soyuz 23	Zudov Rozhdestvensky	Gorbatko Glazkov	*Berezovoi* ??
Soyuz 24	Gorbatko Glazkov	Berezovoi ??	?? ??

The Soyuz 24 back-up flight-engineer has never been identified, and it is therefore assumed that he has not yet made a spaceflight. The support crew assignments are also unconfirmed, and are based on the crewing policies observed on the Salyut 6 and 7 missions *(see sections entitled 'Salyut 6 Crewing' and 'Salyut 7 Crewing' for more details.)* However, there is supporting evidence for Berezovoi's assignment. He has revealed that in April 1975, he found himself at the bottom in a group of crew selections, with no chance of flying. If he was the commander of the support crew for the second mission, as suspected, he would have been in just such a position.

1.12: Salyut 6 Crewing

Introduction

Salyut 6 was the first of the second-generation Salyuts, launched on 29 September 1977. It was equipped with two docking ports, and was the first Salyut which was not wholly civilian or military in nature *(see section entitled 'Military and Civilian Salyut programmes' for more details.)* It was also the first Salyut for which more than two missions were planned. The very first mission failed however, and the effect that this and subsequent events had on the crewing is examined in this section. It should be noted that the crewing tables show only the long-duration crews, and exclude the visiting crews.

Crewing

With the advent of the new Salyut equipped with two docking ports, the Soviets decided to embark on a series of long-duration flights. These would be regularly re-supplied by Progress tankers. However, it was also necessary to provide a means of changing the Soyuz ferry on a regular basis, and it was decided to fly a series of 'visiting missions' which would mainly be used by guest cosmonauts from the Intercosmos countries *(see section entitled 'The Intercosmos programme' for more details.)*

According to the Soviets, only three long-duration flights were originally planned, and it is believed that the crews would have been as follows:

Mission	Prime	Back-up	Support
1	Kovalyonok Ryumin	Romanenko Ivanchenkov	Lyakhov Lebedev
2	Romaneneko Ivanchenkov	Lyakhov Lebedev	Popov Strekalov
3	Lyakhov Lebedev	Popov Strekalov	?? ??

These assignments are speculative, but are based on observations of later flights to Salyut 6, which appeared to reveal a set pattern of rotation for long-duration crews. Each crew seems to have served as a support crew before graduating to back-up status, and finally, prime.

A disruption occurred early in the schedule, when Soyuz 25 failed to dock with the space-station. At one time, it was thought that the Soviets took a policy decision following the failure to no longer fly all-rookie crews. Therefore, the Soyuz 25 back-up crew were split up. Romanenko moved up to prime on Soyuz 26 as planned, and Ivanchenkov was re-assigned to a second back-up duty on that mission. Kovalyonok and Ryumin were 'recycled' – Kovalyonok completing the Soyuz 26 back-up crew, and Ryumin completing the Soyuz 26 support crew. He joined Lyakhov, who therefore had to serve a second support crew assignment.

However, it now seems that the real reason for splitting up the Soyuz 25 back-up crew was a direct result of the decision to insert Grechko into the Soyuz 26 crew. Grechko, who was probably training at the time as back-up commander for the joint Soviet-GDR spaceflight, was apparently chosen because he was a specialist in the Salyut docking system, which was thought to have caused the failure of Soyuz 25. By taking the flight-engineer position on Soyuz 26, he 'bumped' Ivanchenkov, who was switched to the Soyuz 26 back-up crew instead. He in turn 'bumped' Ryumin, who should have stayed with Kovalyonok. Ryumin then 'bumped' Lebedev, and Lebedev 'bumped' Strekalov.

Once Romanenko and Grechko had established that the Salyut 6 docking port was undamaged, Dzhanibekov and Makarov were launched onboard Soyuz 27 to provide the long-duration crew with a fresh spacecraft. Kovalyonok and Ivanchenkov served as the back-ups.

Following the failure of Soyuz 25, it was necessary to fly another mission to Salyut 6 to form the third long-duration flight, Soyuz 26 and Soyuz 29 being the first two. Soyuz 32 was therefore flown, and was crewed by Lyakhov and Ryumin, who should have received three Intercosmos visits *(see section entitled 'The Intercosmos programme' for more details.)*

The revised crewing was therefore as follows:

Mission	**Prime**	**Back-up**	**Support**
Soyuz 26	Romanenko Grechko	Kovalyonok Ivanchenkov	Lyakhov Ryumin
Soyuz 29	Kovalyonok Ivanchenkov	Lyakhov Ryumin	Popov Lebedev
Soyuz 32	Lyakhov Ryumin	Popov Lebedev	Illarionov Strekalov

Unfortunately, the Soyuz 33 mission experienced an engine system failure, and the Soviet-Bulgarian crew had to return to Earth before docking with the Salyut. It was then decided to postpone any further Intercosmos flights until the next long-duration mission, giving engineers time to rectify the problem with the Soyuz engine system.

It was therefore necessary to fly yet another long-duration mission – the fourth – to accomodate three more Intercosmos visits. Soyuz 35 was therefore launched with Popov and Ryumin onboard. Popov should have flown with Valentin Lebedev, but he injured his knee in training shortly before the mission, and Ryumin was asked to stand in.

A final long-duration crew was needed to enable the last two Intercosmos countries to fly, and Kovalyonok and Savinykh were launched onboard Soyuz T-4. They returned to Earth just four days after the end of the final Intercosmos mission, giving further weight to the argument that theirs was a special mission to enable the Intercosmos programme to be concluded.

The final crewing, as flown, was therefore as follows:

Mission	**Prime**	**Back-up**	**Support**
Soyuz 25	Kovalyonok Ryumin	Romanenko *Ivanchenkov*	*Lyakhov* [1] *Lebedev*
Soyuz 26	Romanenko Grechko	Kovalyonok Ivanchenkov	*Lyakhov* *Ryumin*
Soyuz 29	Kovalyonok Ivanchenkov	Lyakhov Ryumin	*Popov* *Lebedev*
Soyuz 32	Lyakhov Ryumin	Popov Lebedev	*Illarionov* [2] *Strekalov*
Soyuz 35	Popov Ryumin	Titov, V [3] Strekalov	*Kovalyonok* *Savinykh*
Soyuz T-4	Kovalyonok Savinykh	Titov, V Strekalov	?? ??

None of the above support crew assignments have been confirmed, and in addition, the identity of the Soyuz 25 back-up flight-engineer has not been revealed. These assignments are therefore speculative and are shown in italics.

Notes

1 It is thought that Dzhanibekov and Andreyev may have been the original Soyuz 25 support crew, but they were replaced when Andreyev became physically disqualified. Dzhanibekov then became available for the Soyuz 27 mission.

2 It is believed that following ASTP, Illarionov trained for a Salyut mission. However, it would appear that at some stage, he became physically disqualified and was replaced. His assignment as support crew commander for Soyuz 32 is very speculative.

3 It is thought that Titov moved up from the support crew to replace Illarionov when he became physically disqualified – see note 2. Kovalyonok then replaced Titov as commander of the support crew.

1.13: Salyut 7 Crewing

Introduction

Salyut 7 was launched on 19 April 1982, and as with Salyut 6, a large number of missions were planned to fly to it. Unfortunately, the space-station suffered a number of serious failures which necessitated a series of repair missions. This section therefore examines the effect that these unexpected missions had on the crewing. It should be noted that the crewing tables show only the long-duration crews, and exclude the visiting crews.

Crewing

As with Salyut 6, there appeared to be a set pattern of rotation for long-duration crews onboard Salyut 7, with each crew serving as a support crew before graduating to back-up status, and finally, prime. A new aspect of the crewing policy took advantage of the Soyuz-T's ability to carry a crew of three. It was intended that ultimately the long-duration crews would consist of three men – considered the optimum number – and that crew handovers would take place. However, these aims were frustrated by the various failures.

It also became apparent that the basic flight crew consisted of the commander and the flight-engineer only. The third crewmember – i.e. the research-cosmonaut – was added only as mission requirements dictated, and it is therefore unlikely that a third crewmember was assigned to the support crew.

It is not possible to say with any certainty how many long-duration flights were originally planned, but it seems reasonable to speculate that the original flight schedule would have been as follows:

Mission	Prime	Back-up	Support
1	Berezovoi Lebedev	Titov Strekalov	Lyakhov Alexandrov
2	Titov Strekalov	Lyakhov Alexandrov	Kizim Solovyov
	Serebrov	Savinykh	??
3	Lyakhov Alexandrov	Kizim Solovyov	Vasyutin Savinykh
	??	??	??
4	Kizim Solovyov	Vasyutin Savinykh	Dzhanibekov Volk
	Atkov	??	??
5	Vasyutin Savinykh	Dzhanibekov Volk	?? ??
	Volkov, A	??	??

The first mission – Soyuz T-5 – was successful, and the crew of Berezovoi and Lebedev set a new duration record and received two visiting crews.

However, a disruption in the schedule occurred with the second mission, when Soyuz T-8 failed to dock. It was the first attempt to place a three-man long-duration crew onboard a Salyut since Soyuz 11, and the cosmonauts were meant to carry out repairs to the solar panels. The back-up crew – believed to be Lyakhov and Alexandrov – minus the research-cosmonaut flew the next mission, Soyuz T-9, and Titov and Strekalov were apparently 'recycled' and flew the following mission in a second attempt to carry out the repairs that they had been trained for. Unfortunately, that mission suffered a launch abort. The Soviets later confirmed that Titov and Strekalov were to have relieved Lyakhov and Alexandrov, thus achieving the first handover of a Salyut from one crew to another.

Kizim and Solovyov are thought to have served as the support crew for Soyuz T-8 and back-up crew for Soyuz T-9. If it had not been for the recycling of Titov and Strekalov, they would have been the prime crew for the original Soyuz T-10, which suffered a launch abort. Instead, they were the back-ups for that mission, and following its failure, were given the replacement Soyuz T-10 assignment. Therefore, apart from the extra back-up crew assignment, Kizim and Solovyov stayed on the original schedule.

They were joined on Soyuz T-10 by a research-cosmonaut – Oleg Atkov – and thus succeeded in becoming the first three-man long-duration crew to occupy a Salyut since 1971.

A further disruption occurred when Salyut 7 suffered a failure in a main fuel line. The decision was taken to fly a repair mission, and Dzhanibekov underwent special training to carry out an EVA to compress and seal the ruptured fuel line. He and Volk were launched onboard Soyuz T-12 together with Svetlana Savitskaya, and it is possible that the two men were also the support crew for Soyuz T-10. Vasyutin and Savinykh were the back-ups on Soyuz T-12, which would indicate that two Salyut long-duration crews were temporarily sidetracked. In the event, Dzhanibekov did not carry out the repairs, as Kizim and Solovyov insisted on doing it themselves.

Following the completion of the Soyuz T-12 mission, the original schedule was resumed, and Vasyutin, Savinykh and A.Volkov were assigned as prime crew on the next mission, Soyuz T-13.

However, yet another disruption occurred when an onboard failure left the unmanned Salyut 7 space-station without power. Another repair mission was necessary, and as the Salyut's docking and rendezvous systems were not operational, a manual docking was called for. As Dzhanibekov had had to carry out a manual docking during the Soyuz T-6 mission, he was the obvious choice to fly the new mission. And as extensive onboard repairs would also be necessary, an experienced flight-engineer was required. Savinykh was given the assignment, and the two men began training at the end of March 1985.

They were launched onboard the replacement Soyuz T-13, Vasyutin having been paired with Georgi Grechko on the back-up crew, and A.Volkov having stood down. Dzhanibekov and Savinykh carried out a manual docking with the Salyut, and managed to restore the space-station to life. Once it was declared operational, the Soviet flight-planners wanted to get back to the original flight schedule. Soyuz T-14 was therefore launched with Vasyutin, Grechko and A.Volkov onboard. After docking with Salyut 7, the two crews worked together for a week, before Dzhanibekov returned to Earth with Grechko.

This was the first time that a partial replacement of crews had taken place, although the capability to do so had existed for many years. It resulted in the Soviets achieving their aim of having the original crew of Vasyutin, Savinykh and A.Volkov onboard the Salyut, thus restoring the original flight schedule. It is clear that the crew fully expected to still be in orbit on 1 January 1986, and it seems likely that Savinykh was intended to set a new duration record. However, in November 1985, Vasyutin became ill, and the mission had to be terminated early.

The Soviets referred to the two crews as the 4th main expedition Stage 1 (Dzhanibekov and Savinykh) and Stage 2 (Vasyutin, Savinykh and A. Volkov.)

The final crewing, as flown, was as follows:

Mission	Prime	Back-up	Support
Soyuz T-5	Berezovoi Lebedev	Titov Strekalov	*Lyakhov*[1] *Alexandrov*
Soyuz T-8	Titov Strekalov	*Lyakhov* *Alexandrov*	*Kizim* *Solovyov*
	Serebrov	*Savinykh*	??
Soyuz T-9	Lyakhov Alexandrov	*Kizim* *Solovyov*	*Vasyutin* *Savinykh*
Launch Abort	Titov Strekalov	Kizim Solovyov	*Vasyutin* *Savinykh*
Soyuz T-10	Kizim Solovyov	Vasyutin Savinykh	*Dzhanibekov* *Volk*
	Atkov	??	??
Soyuz T-13 (cancelled)	Vasyutin Savinykh	*Titov*[2] *Strekalov*[2]	?? ??
	Volkov, A	*Serebrov*[2]	??
Soyuz T-13	Dzhanibekov Savinykh	*Vasyutin* *Grechko*	?? ??
Soyuz T-14	Vasyutin Grechko/Savinykh	*Titov*[2] *Strekalov*[2]	?? ??
	Volkov, A	*Serebrov*[2]	??

None of the above support crew assignments have been confirmed, and in addition most of the back-ups have not been identified. These assignments are therefore speculative and are shown in italics.

Notes

1 Lyakhov must have replaced Dzhanibekov when he was moved to Soyuz T-6, as it is known that Dzhanibekov and Alexandrov trained together for a long-duration flight.

2 This is highly speculative – see biography.

1.14: ASTP Crewing

Introduction

The Apollo-Soyuz Test Project (ASTP) was the first joint spaceflight between the Soviet Union and the United States of America. It involved the docking of an American Apollo with a Soviet Soyuz spacecraft, and whilst it was born in, and was a symbol of, a period of detente between the two countries, many useful scientific experiments were carried out. Above all, it was a clear demonstration that despite the enormous technical, political, social and linguistic differences, the two countries could work together in space.

Background

The joint mission called for the Soyuz to be launched first, followed by the Apollo, complete with a Docking Module. Such was the importance that the Soviets attached to the success of the mission that they had a complete back-up Soyuz spacecraft, launcher and crew ready and waiting on a separate launch pad. If a major problem had prevented the launching of the prime crew, the second crew and spacecraft would have been launched instead.

In addition, a complete dress rehearsal of the flight was carried out by the Soyuz 16 mission, launched on 2 December 1974. Anatoli Filipchenko and Nikolai Rukavishnikov, who had been assigned as the prime crew for the reserve spacecraft, carried out a complete rehearsal of the ASTP Soyuz flight programme, including important docking tests. They returned to Earth after a flight lasting 5 days 22 hours and 24 minutes.

The actual ASTP spacecraft, Soyuz 19, was launched on 15 July 1975, and the crew of Alexei Leonov and Valery Kubasov spent two days in orbit preparing for the historic docking with the American Apollo spacecraft. The two craft remained docked for nearly two days, during which the crews carried out joint experiments and transfers to each others spacecraft. The two craft then separated, and the Soyuz 19 crew returned to Earth after a flight lasting 5 days 22 hours and 31 minutes.

Following the successful launch of the prime crew, the back-up commanders, Vladimir Dzhanibekov and Yuri Romanenko, served as CapComs for the mission. The Chief CapCom was Georgi Shonin.

Crewing

Mission	Prime	Back-up	Support
Soyuz 16	Filipchenko Rukavishnikov	Romanenko Ivanchenkov	
Soyuz 19	Leonov Kubasov	Dzhanibekov Andreyev	
Reserve ASTP Soyuz	Filipchenko Rukavishnikov	Romanenko Ivanchenkov	
Apollo	Stafford Brand Slayton	Bean Evans Lousma	Truly Overmyer Crippen Bobko

The American crews were named on 30 January 1973. During the preparations for ASTP, the support crews stood in for the prime crews in a number of time-consuming but critical activities, such as mission-planning and lengthy manned tests of the flight hardware. During the flight, Truly, Crippen and Bobko served as CapComs in MCC at Houston. Overmyer, who worked extensively with the Soviets on mission planning and crew training, was one of the technical advisers at the Soviet FCC.

The Soviet crews were named on 24 May 1973, to co-incide with the opening of the Paris Air Show. This public announcement of crews was a first for the Soviets, who in the past had never identified crew assignments in advance.

Until the announcement, Vladimir Shatalov and Alexei Yeliseyev had been identifying themselves to their Western colleagues as the designated crewmen for ASTP. It is believed that Leonov and Kubasov were given the assignment when their planned mission to a Salyut was cancelled following the Cosmos 557 failure in May 1973. *(see section entitled 'Salyut 2/Cosmos 557 Crewing' for more details.)*

The four prime Soviet crewmen were all experienced cosmonauts, but the four back-ups – Boris Andreyev, Vladimir Dzhanibekov, Alexander Ivanchenkov and Yuri Romanenko – were rookie cosmonauts, selected in 1970. This was another first for the Soviets, as they had never previously identified a cosmonaut before a first spaceflight. Three of these men have gone on to make spaceflights, but Andreyev seems to have become disqualified, as does another cosmonaut identified at the time of ASTP – Valery Illarionov – who served as a CapCom at MCC in Houston.

1.15: Intercosmos Crewing

Introduction

The Soviets adopted a clearly defined crewing policy for the commanders of the Intercosmos missions, and this section examines that policy, together with the various anomalies which were later observed with the research-cosmonaut assignments. *(For details concerning order of flight etc. see section entitled 'The Intercosmos programme'.)*

Crewing Policy

With the launch of the fourth Intercosmos mission, Soyuz 33, it became apparent that there was a set pattern of rotation for the Intercosmos crew commanders. Having served as a back-up commander, a cosmonaut would then fly as commander of the third flight following e.g. Rukavishnikov was back-up on Soyuz 28, and flew as commander of Soyuz 33.

This crewing policy only applied to the original Intercosmos programme, conducted using the Salyut 6 space-station. With the addition of the French and Indian missions, conducted using Salyut 7, no such policy existed, and those missions are only included here for the sake of completeness.

Anomalies

1: Commanders

The only exception to the established crewing policy for commanders was Valery Bykovsky. He served as back-up on Soyuz 37, and should therefore have flown as commander of Soyuz 40. In the event however, Bykovsky apparently became disqualified, and the position was offered to Yevgeny Khrunov, the next in line under the crewing policy. When he declined, the back-up crew of Leonid Popov and Dumitru Prunariu flew the mission.

There is some evidence to suggest that Georgi Grechko was the original back-up commander for the joint Soviet-GDR spaceflight. He was moved into the Soyuz 26 crew following the Soyuz 25 failure, as he was a specialist in the Salyut docking mechanism, and his place on the GDR mission was taken by Viktor Gorbatko.

2: Research-cosmonauts

Confirmation of the Soyuz 40 switch came when it was revealed that Dumitru Dediu was in fact chosen by the Soviet Union to be the first Romanian in space. He was named as such in a United Press report issued in Moscow shortly before the mission began. It therefore seems that Bykovsky and Dediu should have flown, but when Bykovsky became disqualified, the back-up crew of Popov and Dumitru Prunariu took the prime slot. Dediu was moved to back-up, and Yuri Romanenko became the new back-up commander.

Another unexpected change in the crewing became apparent when it was discovered that Zenon Jankowski, the back-up on Soyuz 30, was in fact, chosen by the Soviet Union to be the first Pole in space. The Polish Government therefore printed two stamps honouring him as part of the 1978 Intercosmos programme. However, the day before the flight, the Polish Philatelic Agency, Ars Polona, received word that Miroslaw Hermaszewski was to be the first Polish cosmonaut. New stamps were immediately printed, and the incorrect stamps were destroyed, but some apparently escaped the shredding machines.

The final anomaly concerned the two Indian cosmonauts. Photographs were released showing Malhotra training with Yuri Malyshev and Nikolai Rukavishnikov, and Sharma with Anatoli Berezovoi and Georgi Grechko. It was assumed that these were the crews which would make the actual flight to Salyut 7, with the Malhotra crew as the prime choice, as Malyshev had originally been scheduled to command the joint Soviet-French mission before he was stood down.

However, when the crews for the joint flight were officially named, the prime crew consisted of Malyshev, Rukavishnikov (subsequently replaced by Gennady Strekalov) and Sharma; the reserve crew was Berezovoi, Grechko and Malhotra. Therefore, for some reason Malhotra had been replaced as prime by Sharma.

Crewing

Mission:-	Country:-	Prime:-	Back-up:-
Soyuz 28	Czechoslovakia	Gubarev Remek	Rukavishnikov Pelčak
Soyuz 30	Poland	Klimuk Hermaszewski	Kubasov Jankowski
Soyuz 31	GDR	Bykovsky Jahn	Gorbatko Kollner
Soyuz 33	Bulgaria	Rukavishnikov Ivanov	Romanenko Alexandrov
Soyuz 36	Hungary	Kubasov Farkas	Dzhanibekov Magyari
Soyuz 37	Vietnam	Gorbatko Tuan	Bykovsky Liem
Soyuz 38	Cuba	Romanenko Tamayo-Méndez	Khrunov Lopez-Falcon
Soyuz 39	Mongolia	Dzhanibekov Gurragcha	Lyakhov Ganzorig
Soyuz 40	Romania	Popov Prunariu	Romanenko Dediu
Salyut 7			
Soyuz T-6	France	Dzhanibekov [1] Ivanchenkov Chrétien	Kizim Solovyov Baudry
Soyuz T-11	India	Malyshev Strekalov [2] Sharma	Berezovoi Grechko Malhotra

Notes

1 Dzhanibekov replaced Yuri Malyshev because of personal incompatibility.

2 Strekalov replaced Nikolai Rukavishnikov when he became ill.

1.16: The Intercosmos Programme

Introduction

The Intercosmos programme consists of co-operative ventures, usually unmanned, between the Soviet-Bloc countries. However in 1976, this was broadened to include manned spaceflights by representatives of each country onboard a Soviet spacecraft. This section details the rationale behind the Intercosmos flights, the Intercosmos organization, the agreements governing the flights, selection dates, training, mission plans, order of first flight, the reasons for flight order and the selection procedures followed by each country. It should be read in conjunction with the section entitled 'Intercosmos Crewing.'

As well as the Intercosmos countries, France and India have also taken part in joint spaceflights with the Soviets, and these missions are also included in this section.

All the missions except the Bulgarian one were successful. Following the Soyuz 33 failure, the Bulgarian cosmonaut, Georgi Ivanov, was not given a further flight opportunity.

The Rationale behind the Intercosmos Flights

With the advent of the new generation of Salyuts equipped with two docking ports, (the first being Salyut 6) the Soviets planned to embark on a series of long-duration flights. These would be regularly re-supplied by Progress tankers.

However, there would be a requirement to conduct 'visiting missions' during which the visiting crew would leave their Soyuz docked to the Salyut whilst returning to Earth in the spacecraft originally used by the long-duration crew. This exchange of ferry craft was necessary as the Soyuz had only been proven to be capable of being safely re-activated after a 90-day period of dormancy. Many of the planned long-duration flights would exceed this period.

Having devised the method of providing a fresh spacecraft, the Soviet planners set about making the best use of the planned visits. It was decided that they could be used to provide flight opportunities to Intercosmos member countries.

Whilst the flights were mainly politically motivated, the Intercosmos cosmonauts did manage to carry out a broad range of experiments designed and constructed by their own country's scientists. And although the Soviets reaped large propaganda dividends from the programme, it seems that they also gained greatly in the technical and scientific fields as a result of using experiments and equipment designed by the other countries.

The Intercosmos Organization

Intercosmos, the Soviet acronym for the Council for International Co-operation in the Exploration and use of Outer Space, was established in 1966 within the Soviet Academy of Sciences. Its general task was to co-ordinate the work of the various ministries and departments involved in international programmes on space research, with the primary function of overseeing and co-ordinating Soviet space co-operation with Eastern European and other Communist countries.

Intercosmos had its beginnings in November 1965, but it was not until April 1967, following a series of meetings and conferences, that the Intercosmos programme of Communist space co-operation was formally adopted.

National co-ordinating bodies called National Agencies were established in the nine member countries: Bulgaria, Czechoslovakia, Cuba, the German Democratic Republic, Hungary, Mongolia, Poland, Romania and the Soviet Union. Annual conferences are held to review progress and formulate new long term tasks for co-operation in space.

Five major areas of studies are undertaken: Space physics, space meteorology, space communications, space biology and medicine, and the study of the natural environment by means of space techniques.

Intercosmos does not have a common budget – each country simply finances that space experiment in which it has an interest. The Soviet Union however, has made available its space technology to the other member countries free of charge.

Agreements concerning non-Soviet cosmonauts

Talks were held in Moscow in July 1976 amongst delegations from the Intercosmos member countries after the Soviets had proposed the participation of Bulgarian, Cuban, Czechoslovakian, East German, Hungarian, Mongolian, Polish and Romanian cosmonauts in Soviet spacecraft and space-stations.

The delegations examined questions concerning the selection of the candidate cosmonauts and their training in the Soviet Union, and adopted agreed recommendations on all the questions discussed.

The meeting ended with the signing in Moscow by all the member countries on 13 July 1976 of an agreement which planned to promote 'the further development of international co-operation in the exploration and uses of outer space for peaceful purposes.'

A more specific agreement was signed in Moscow on 14 September 1976, under which the Intercosmos cosmonauts would fly onboard Soviet spacecraft in the period 1978 – 83.

Commenting on the agreement, Vladimir Shatalov pointed out that the timing of such flights depended on space vehicles and equipment being ready and on the preparedness of the crews. 'It will take us about a year and a half to train the crews, provided, of course, that there is no language barrier' he said.

Shatalov explained that the crew commander would always be a Soviet cosmonaut, while the flight-engineer and research-engineer would come from other Communist countries.

Crew members, he continued, would be selected in the light of the nature and scope of the experiments planned for the various flights. It had been found adviseable, in the light of experience gained in both the US and the Soviet Union – and also in view of the size and training period of the first international crews – to train as crew members experienced jet-pilots who were well grounded in engineering. That did not rule out the eventual utilisation of specialists in other fields.

In the period April – June 1979, three more countries were invited to take part in joint spaceflights.

In the course of President Giscard d'Estaing's visit to the Soviet Union in April 1979, Leonid Brezhnev proposed that a French cosmonaut take part in a Soviet spaceflight. This proposal was duly considered and accepted.

In May 1979, it was reported that Vietnam had joined the Intercosmos programme, and

Vietnamese cosmonauts were already in training at Star Town. And in June 1979, it was announced that as an extension of its space co-operation programme with India, the Soviet Union had offered India a joint spaceflight.

A new series of joint flights was proposed in 1985. In April, it was reported that the Soviet Union had invited Japan to participate in a manned flight on a Soviet spacecraft, and in October, agreement was reached on a second French spaceflight, involving an extended stay onboard a Salyut. And finally, in late 1985, it was announced that two Syrian cosmonauts were in training for a joint Soviet-Syrian spaceflight.

Selection Groups

December 1976	Czechoslovakia	Remek and Pelčak
	GDR	Jähn and Köllner
	Poland	Hermaszewski and Jankowski
March 1978	Bulgaria	Ivanov and Alexandrov
	Cuba	Tamayo-Méndez and Lopez-Falcon
	Hungary	Farkas and Magyari
	Mongolia	Gurragcha and Ganzorig
	Romania	Prunariu and Dediu
April 1979	Vietnam	Liem and Tuan
12 June 1980	France	Baudry and Chrétien
September 1982	India	Malhotra and Sharma
late 1985	Syria	Fares and Habib

Intercosmos Training

The training of the cosmonauts seemed to be intensive in every respect. The first stage included theoretical studies combined with flight experience in jet planes, physical exercises, simulation of zero-G flights, and practicing elementary recovery procedures, including splashdowns and retrieval by helicopters from thick forests, mountains and other difficult terrain. The second stage consisted of mastering the Soyuz spacecraft together with the specific flight programme for the planned mission onboard the Salyut space-station.

Mission Plans

The Soviets have announced that only three long-duration missions were originally planned for Salyut 6 *(see section entitled 'Salyut 6 Crewing' for more details.)* It would therefore seem that all of the eight original Intercosmos missions were planned to take place during these three flights. However, the Soyuz 25 failure upset the plan, and only one Intercosmos mission was flown to visit the first long-duration crew. It would therefore appear that another long-duration mission was added, and the following plan devised:

Crew No.	Cosmonauts	Intercosmos Country
1	Romanenko/Grechko [1]	Czechoslovakia
2	Kovalyonok/Ivanchenkov	Poland, GDR.
3	Lyakhov/Ryumin	Bulgaria, Hungary, Cuba.
4	Popov/Ryumin [2]	Mongolia, Romania.

Notes

1 The first crew should have been Kovalyonok and Ryumin, but Soyuz 25 failed to dock.

2 Ryumin replaced Lebedev, who was injured shortly before the flight.

The first three flights went as planned, but then in April 1979, the Soyuz 33 mission experienced an engine system failure, and the Soviet-Bulgarian crew had to return to Earth before docking with the Salyut. It was then decided to postpone any further Intercosmos flights until the next long-duration mission, giving engineers time to rectify the problem with the Soyuz engine system.

Reports from Hungary on 2 June 1979 indicated that the Hungarian cosmonaut had been scheduled to be launched on 5 June to dock with Salyut 6, but the mission had been postponed because of concern about the Soyuz engines. Technicians had apparently identified the problem on Soyuz 33, but did not want to commit a crew to a Soyuz until an unmanned test could be completed. Thus, the unmanned Soyuz 34 was launched on 6 June. After 2 days of independent flight to test out the modified engine, Soyuz 34 docked with the Salyut.

Five days later, Soyuz 32 undocked and carried 180 kg (397 lbs) of materials down to Earth. Since the Soyuz 33 failure had precluded visiting missions, there was no other method of getting the experiment results back except with the main crew.

The spacecraft landed after a record 109 days in space, and was recovered on the same day as the planned Hungarian flight was probably scheduled to end, lending support to the theory that cosmonauts from both Hungary and Cuba were to have visited the Lyakhov/Ryumin crew. Analysis of the launch/landing windows gives an indication of the planned timetable:

Mission	Country	Launch	Landing
Soyuz 34	Hungary	5 June	13 June
Soyuz 35	Cuba	15 June	23 June

The Hungarian and Cuban flights were therefore delayed by a year. In addition, Vietnam had been admitted to the Intercosmos programme in May 1979, and its flight was inserted in between that of Hungary and Cuba. This was because of the alphabetical system used to determine the order of flight (see Reasons for Flight Order below) and possibly because of the propagandist effect during the 1980 Moscow Olympic Games.

To enable the last two countries – Mongolia and Romania – to fly, another long-duration crew was needed, and Kovalyonok and Savinykh were given the assignment. The Romanian mission had to go to Salyut 6 because of the ASTRO experiment, which used the ASTRO-1 instrument fixed outside the Salyut 6 station, and ASTRO-2 mounted on the exterior of the Soviet-Romanian Soyuz 40 spacecraft. Kovalyonok and Savinykh returned to Earth just four days after the end of the Romanian mission, giving further weight to the argument that theirs was a special mission to enable the Intercosmos programme to be concluded.

The revised flight order thus became:

Crew No.	Cosmonauts	Intercosmos Country
1	Romanenko/Grechko	Czechoslovakia
2	Kovalyonok/Ivanchenkov	Poland, GDR.
3	Lyakhov/Ryumin	Bulgaria.
4	Popov/Ryumin	Hungary, Vietnam, Cuba.
5	Kovalyonok/Savinykh	Mongolia, Romania.

Intercosmos countries in order of flight

No.	Country	Cosmonaut	Launch Date
1	Czechoslovakia	Remek	2 March 1978
2	Poland	Hermaszewski	27 June 1978
3	GDR	Jähn	26 August 1978
4	Bulgaria	Ivanov	10 April 1979
5	Hungary	Farkas	26 May 1980
6	Vietnam	Tuan	23 July 1980
7	Cuba	Tamayo-Méndez	18 September 1980
8	Mongolia	Gurragcha	22 March 1981
9	Romania	Prunariu	14 May 1981
10	France	Chrétien	24 June 1982
11	India	Sharma	3 April 1984

Reasons for Flight Order

When the decision was taken to allow Soviet-Bloc countries to participate in the Soviet space programme, the order in which each country would fly had to be decided.

The Intercosmos cosmonauts were selected in two main groups – 1976 and 1978 – and no explanation has been given as to why the countries were split in this way. Furthermore, there is no apparent explanation for what determined the order of the flights by the first group. The first three countries to fly were:

Czechoslovakia

Poland

German Democratic Republic

However, a clear pattern has been established for the second group. The cosmonauts selected in 1978, together with the Vietnamese cosmonaut selected in 1979, were launched in strictly alphabetical order. The following shows each country in alphabetical order using the cyrillic alphabet, and Russian language names:-

Bolgaria (Bulgaria)

Vengria (Hungary)

Vietnam (Vietnam)

Kuba (Cuba)

Mongolia (Mongolia)

Rumynia (Rumania)

It has been pointed out that the choice of the politically neutral alphabetical system for determining the order of flight was somewhat ironic in a programme which existed largely for political propaganda purposes. However, it had the advantage of not forcing the Soviet Union to try and judge the loyalty of its allies in public. In addition, as some political observers pointed out, the scheme still allowed the Soviet Union to make Romania, with whom relations had been cool at the end of the 1970's, to wait until last.

As can be seen above, the Vietnamese cosmonaut was launched in accordance with the alphabetical system. However, as the flight came during the 1980 Moscow Olympics, it was seen in some quarters as an attempt by the Soviet Union to impress the Third World. It should be noted that if it had not been for the Soyuz 33 failure, Vietnam would **not** have flown in order, as the Vietnamese cosmonauts were selected in April 1979, much later than the other Intercosmos cosmonauts, and would not have completed their training in time to fly after Hungary. (see Mission Plans above.)

Selection Procedures

Bulgaria

There were apparently six Bulgarian cosmonaut candidates:

Alexander Alexandrov, Chavdar Djurov, Georgi Iovchev, Georgi Ivanov, Ivan Nakov, and Kyril Radev.

Georgi Ivanov flew the Soyuz 33 mission and Alexander Alexandrov served as the back-up. Brief biographical details have been released on the remaining candidates:

Captain-Engineer Chavdar Djurov

The son of the Bulgarian Minister of National Defence, General Dobri Djurov, he was apparently an excellent parachutist, and held the World record for the highest altitude jump made at night – 13,000 metres. He was killed in a L-29 jet-fighter crash when he was flying aerobatics and failed to pull out of a loop. This was said to be in 1972 – but this is obviously in error as he couldn't then have been a cosmonaut-candidate. His wife was a stewardess in the Bulgarian Balkan civil airline, and they had a daughter, aged 8.

Major-Engineer Georgi Iovchev

A 35 year old fighter pilot and parachutist with a Master of Science degree, fluent in English and Russian. He is now a flying instructor at the Headquarters of the Bulgarian Air Force.

Captain-Engineer Ivan Nakov

A 31 year old fighter pilot who graduated from a Higher Air Force Academy in 1972.

Lieutenant-Colonel Kyril Radev

A 41 year old fighter pilot, and apparently one of the best Bulgarian supersonic pilots flying MiG 21's. He is now Commander of a squadron of jet-fighters made up of MiG 17, MiG 21 and MiG 23's.

Note

All ages and career details are as given in the 1981 source.

Cuba

A large number of Cuban Air Force pilots were picked for aptitude tests and medical examinations. Nine finalists were chosen, of which four were sent to the Soviet Union in January 1978. They spent the last ten days of January and all of February undergoing numerous tests before returning to Cuba.

The two finalists were Arnaldo Tamayo-Méndez and José Lopez-Falcon, who learnt of their selection on 1 March 1978 and began their training at the Yuri Gagarin Cosmonaut Training Centre on 22 March 1978.

Czechoslovakia

After an initial screening of the service records of all Czechoslovakian Air Force pilots, a few dozen men were picked for further tests in July 1976. They then underwent a series of medical and psychological tests at the Czech Military Aviation Institute.

By the end of October, four finalists were chosen, and they were sent for further tests at the Yuri Gagarin Cosmonaut Training Centre. They arrived at Star Town on 10 November, and underwent two weeks of tests before returning to Czechoslovakia.

The next day, the two finalists were named – Vladimir Remek and Oldrich Pelčak – and they returned to Star Town on 6 December 1976 to begin their training.

Remek began joint training with Alexei Gubarev in August 1977, and at the end of February 1978, Remek and Pelčak, together with the Polish and GDR cosmonauts, sat their final exams and tests at the Yuri Gagarin Cosmonaut Training Centre. The results were extremely close, but eventually, Remek was selected to be the first representative of the Intercosmos countries to fly in space.

France

In the course of President Giscard d'Estaing's visit to the Soviet Union in April 1979, Leonid Brezhnev proposed that a French cosmonaut take part in a Soviet spaceflight. This proposal was duly considered and accepted.

The French space agency CNES (Centre National d'Etudes Spatiales) began a selection process in September 1979. By 15 November 1979, CNES had received 413 applications. Of these, 193 (including 26 women) were called up for the first phase of the selection, which involved a medical and an overall assessment, which looked at their general competence, familiarity with scientific equipment, linguistic capacity, and sporting abilities. Following this first phase, 72 candidates remained.

The second phase, in December 1979 and January 1980, consisted of a preliminary examination; vestibular tests; a complete medical examination lasting a week; a psychological examination lasting 3 days; a psychiatric examination; and a special set of physiological tests lasting a week.

At the end of this second phase, 7 candidates were left, including 2 women.

In parallel with the special physiological tests, the third phase of the selection began in January and February 1980 – tests of general competence and scientific aptitude. It consisted of three series of interviews for judging respectively:

1) the scientific aptitude of the candidates – interview conducted by scientific experts;

2) their linguistic aptitude – interview conducted by linguistic experts;

3) their personal presentation – interview conducted by CNES experts.

At this stage, 5 candidates remained – four men and one woman. They were Patrick Baudry, Jean-Loup Chrétien, Gerard Juin – an Air France Airbus pilot, Jean-Pierre Job – an Air Force pilot with the Commandant Squadron at Normandie-Nieman, and Francoise Varnier. The final stage consisted of a parachute jump and a flight on an Epsilon at the Test Flight Centre at Brétigny. A parachute accident eliminated Varnier, the final woman candidate.

In March 1980, the team went to Moscow for 3 months studies of the Russian language.

Shortly after their return, the two final candidates – Baudry and Chrétien – were named on 12 June 1980. They commenced their training at Star Town on 7 September 1980.

Chrétien eventually flew the Soyuz T-6 mission, and Baudry served as back-up. Baudry then flew on the American Shuttle mission 51-G, with Chrétien as back-up. The two men are on detachment to CNES from the Air Force until at least 1989, and will remain available for future flights with the Shuttle (NASA), Salyut (Soviet Union) or Spacelab (ESA).

In 1985, France opened a recruitment programme to expand its astronaut corps, and CNES's goal was to create a reserve of crewmembers available for future flights. More than 600 applications were received and seven new candidates were selected in September 1985. They were Jean Francois Clervoy, Claudie Deshays, Jean Jacques Favier, Jean-Pierre Haignere, Frédéric Patat, Michel Tognini and Michel Viso.

The French scientific community had been seeking a flight of approximately 2 months on a Salyut as a follow-up to Chrétien's mission, and agreement on a second joint flight was finally reached in October 1985 during Mikhail Gorbachyov's visit to France.

GDR

In late July 1976, the Vice-Minister of Defence, Commander-in-Chief of the GDR Air Force, Lieutenant-General Wolfgang Reinhold personally interviewed hundreds of the best East German Air Force pilots. Each suitable candidate was told that volunteers were needed for selection as cosmonaut candidates, and were given 2 days to make up their mind.

At the end of September, the 30 best candidates underwent further tests. Selection criteria had included a minimum age of 35, a maximum age of 45, over 1000 hours in jet planes, an excellent military service record and academic achievement (graduation from a military academy, preferably the Yuri Gagarin Air Force Academy in Moscow.) The men underwent medical and psychological tests and were educated in physics, mathematics, Russian and sports, at the Medical Institute for Pilots.

About a month later, four finalists were chosen for further selection tests at the Yuri Gagarin Cosmonaut Training Centre. The four men – Rolf Berger, Eberhard Golbs, Sigmund Jähn and Eberhard Köllner – all Lieutenant-Colonels – arrived at Star Town on 10 November, and underwent two weeks of tests before returning to the GDR.

The two finalists – Jähn and Köllner – were named the following day. They returned to Star Town on 6 December 1976, and began training the next day.

Jähn eventually flew the Soyuz 31 mission, and Köllner served as back-up.

Hungary

After an initial screening of the service records of all Hungarian Air Force pilots, over a thousand who met the criteria – which included a maximum height of 5ft 10in (180cm), a maximum age of 35, and a proven ability to learn the Russian language – were invited to take part in further selection tests.

By the end of September 1976, the number of candidates had been reduced to about 20 after further screening of all records. These candidates then had to undergo a harsh selection process, lasting two weeks, at the Scientific Institute for Aviation at Kecskeme. The selection procedures consisted mainly of psychological and medical tests, including tests of willpower, intelligence, temperament, character, self-control, and physical endurance. The medical tests were supervised by physician Colonel Dr Janos Hideg, while the psychological ones were led by Colonel Dr Jozef Szabo.

By the end of the year four finalists had been chosen, and they were sent to Star Town on 21 January 1978. They then underwent a five week selection process at the Yuri Gagarin Cosmonaut Training Centre before returning to Hungary on 28 February.

The results were announced the next day, and on 13 March, the two finalists – Bertalan Farkas and Bela Magyari – were presented to the Hungarian Press.

They returned to Star Town on 20 March, and began their training the next day. Farkas eventually flew the Soyuz 36 mission, and Magyari served as back-up.

India

A joint Soviet-Indian spaceflight was first suggested during Yuri Gagarin's visit to India in 1961, and during Mr Kosygin's visit during the Janata regime. In 1978, the idea was raised with Prime Minister Desai and again in June 1979, with Mrs Gandhi. Agreement was finally reached during President Brezhnev's visit to India at the end of 1980, after Mrs Gandhi was returned to power. After further discussions, Mrs Gandhi announced in the Indian Parliament in March 1981 that India had accepted the offer.

In October 1981, it was announced that 10 candidate-cosmonauts had been chosen from the 150 original applicants, and that they were to undergo additional tests under a system devised for the training of Soviet cosmonauts. All 10 were officers of the Indian Air Force, and had undergone examinations and preliminary training in various Air Force centres, including the Institute of Aviation Medicine in Bangalore.

Six out of the 10 were then sent to Star Town where they were submitted to further selection procedures, after which two finalists were chosen.

On 20 September 1982, before leaving for a visit to the Soviet Union, Mrs Gandhi told the Indian Parliament that two pilots had been sent to Moscow that month to train for the flight. When she visited Star Town on 23 September 1982, she was able to meet the two pilots – Rakesh Sharma and Ravish Malhotra – who had arrived on 19 September 1982.

Mongolia

The two Mongolian cosmonaut-candidates were Jugderdemidiin Gurragcha and Majdarzavyn Ganzorig. Gurragcha flew the Soyuz 39 mission, and Ganzorig served as back-up.

No other details were released concerning selection procedures, and no other cosmonaut-candidates have been identified.

Poland

The Polish Military Institute for Air Force Medicine and Psychology worked out the basic criteria for the selection of Polish cosmonaut-candidates. The requirements were unspecified university studies, experience as a pilot and knowledge of the subjects of research.

The Polish specialists concentrated their selection on military pilots, and from the existing medical documents, records and personal data available on military pilots, a few thousand candidates were picked out and subjected to very stringent and exhaustive tests.

Several dozen candidates remained, and from these, a group of 18 candidates, all military pilots, were selected in August 1976.

This group then underwent a training and selection programme, and from this came five successful candidates – Major Andrzej Bugala; Major Henryk Halka; Miroslaw Hermaszewski; Zenon Jankowski and Pilot-engineer Tadeusz Kuziora. These five started a two week series of tests on 10 November 1976 at the Yuri Gagarin Cosmonaut Training Centre in Star City. The Main Medical Commission of the Soviet Union then pronounced all five candidates fit to take part in the space programme. Based on the Commission's findings, the Polish experts chose two men to be sent for cosmonaut training.

The two successful candidates were Miroslaw Hermaszewski and Zenon Jankowski. Hermaszewski flew the Soyuz 30 mission, and Jankowski served as back-up.

However, according to unofficial philatelic information from Poland, Zenon Jankowski was in fact, chosen by the Soviet Union to be the first Pole in space. The Polish Government therefore printed two stamps honouring cosmonaut Jankowski as part of the 1978 Intercosmos programme.

However, the day before the flight, the Polish Philatelic Agency, Ars Polona, received word that Miroslaw Hermaszewski was to be the first Polish cosmonaut. New stamps were immediately printed, and the incorrect stamps were destroyed, but some apparently escaped the shredding machines.

Romania

The two Romanian cosmonaut-candidates were Dumitru Prunariu and Dumitru Dediu. Prunariu flew the Soyuz 40 mission, and Dediu served as back-up.

However, according to unofficial information, Dediu was in fact, chosen by the Soviet Union to be the first Romanian in space. He was named as such in a United Press report issued in Moscow shortly before the mission began. It therefore seems that Valery Bykovsky and Dediu should have flown, but when Bykovsky became disqualified, the back-up crew of Popov and Prunariu took the prime slot. Dediu was moved to back-up, and Yuri Romanenko became the new back-up commander. This switch of crews was also reported by United Press.

Syria

In late 1985, it was announced that Syria had been invited by the Soviet Union to participate in a joint spaceflight, and had accepted. Shortly afterwards it was reported that the two Syrian cosmonauts – Mohammad Al Fares and Munir Habib – had begun training at the Yuri Gagarin Cosmonaut Training Centre.

No other details were released concerning selection procedures, and no other cosmonaut-candidates have been identified.

Vietnam

In May 1979, Vietnam became a member of the Intercosmos programme, and as such was immediately invited by the Soviet Union to participate in a joint spaceflight, in the same way as the other member countries of Intercosmos. It was then revealed that Vietnamese cosmonauts were already undergoing training at the Yuri Gagarin Cosmonaut Training Centre.

Pham Tuan and Bui Thanh Liem were those cosmonauts, and they had begun their training in April 1979.

No other details were released concerning selection procedures, and no other cosmonaut-candidates have been identified.

1.17: The Zond Programme

Introduction

Zond is Russian for Probe, and was the name given to a series of unmanned spacecraft sent around the Moon in the late 1960's and early 1970's. They were precursors for an intended series of manned flights which were delayed by technical problems, and later abandoned following the success of the American Apollo missions. It is beyond the scope of this book to give a detailed analysis of what was, by its nature, a very complex part of the Soviet manned spaceflight programme. However, an overview is included because of the importance of the programme and the impact it had on cosmonaut assignments and training.

Therefore, this section covers the background to the Zond programme, the unmanned flights, the Zond Mission Profile and Crewing Assignments.

Background

Statements by Soviet officials and cosmonauts during the 'Moon Race' period of the 1960's certainly gave the impression that the Soviet Union was trying to beat America in becoming the first country to send a man to the Moon.

In 1965, after the success of the first two flights of Voskhod, it is said that Sergei Korolyov decided to use a modified version of the new Soyuz spacecraft for a circumlunar flight. It was initially planned for November 1967, to co-incide with the 50th anniversary of the October Revolution. However, following Korolyov's death, his successor, Vasili Mishin, postponed the mission to December 1968, and Pavel Belyayev started training for the flight in late 1967.

Five unmanned Zond flights conducted in 1968-70 were officially identified as having been engineering tests related to the development of a manned spacecraft to go to the Moon, and a 1971 Cosmos flight has been identified as having been a test of a 'lunar cabin.' No more Zond flights have been made since then, and the programme appears to have been abandoned following the success of the American Apollo programme. The Soviets then began a policy of playing down the importance of manned lunar exploration and stressed the benefits of unmanned flights to the Moon i.e. their Luna and Lunokhod missions. This was of course in direct contradiction to their earlier statements in which they repeatedly claimed that Soviet cosmonauts would be the first to land on the Moon.

However, in 1981, the existence of the Soviet manned lunar programme was officially confirmed by Vladimir Sevastyanov, when he named the cosmonauts involved. They had begun training in 1967. (See Crewing Assignments below).

It seems that the Soviets were prevented from going to the Moon by their apparent inability to successfully build a suitably powerful launch vehicle. This booster – the so-called G-vehicle – would have been the equivalent of the American Saturn V rocket.

Although it appeared that the Soviets abandoned their manned lunar programme once they were beaten by the Americans, it is interesting to note that the Cosmos flights related to the development of a 'lunar cabin' were launched two years after the 1969 landing of Apollo 11. In addition, it is known that lunar landers were still around and being developed in 1977.

Unmanned Zond Flights

Zond 1-3

These were non-lunar missions, and were flown to other planetary targets.

Zond 4

Zond 4, launched on 2 March 1968, was a test-flight sent in the opposite direction from the location of the Moon. It was probably intended to go as far as the Moon's orbit.

A drawing released in 1971 showed Zond 4 to be virtually identical in external appearance to a Soyuz spacecraft without the orbital module.

Zond 5

Zond 5 was launched on 14 September 1968. On this occasion the spacecraft was directed towards the Moon, and it made mid-course corrections to enable it to swing around the Moon at a distance of 1211 miles (1950 km) and then return to Earth.

The Zond spacecraft was described as having two compartments. The first was a recoverable cabin with a heat-regulating system, power supply, parachute packs, scientific instruments, and radio communications equipment. The second was a service module with large solar panels, a radio telemetry system, control equipment, orientation and stabilization systems, batteries, and heat regulation systems.

Zond 5 returned to Earth after 7 days in space, and splashed down in the Indian Ocean at night, having approached the Earth over the South Pole.

Zond 6

Zond 6 was launched on 10 November 1968, and was basically a repeat of the Zond 5 mission. One notable difference was that it performed a skip-return trajectory.

Using aerodynamic braking, the spacecraft was slowed from 6.83 miles (11 km) per second to 4.72 miles (7.6 km) per second. It was then oriented so that it developed lift and skipped outside the atmosphere, before making a second re-entry and a controlled landing in the Soviet Union.

Soviet officials explained that a direct ballistic approach would cause a G-force overload for a human crew, and for this reason the South Pole approach was the only one practical for returning Zond payloads to the Soviet Union. Academician Petrov stated that the G-load for Zond 5 reached 10-16 G's, and implied that for Zond 6 it was closer to that encountered with a manned Soyuz mission (3-4 G's.)

Following the Zond 6 mission, the Soviets formally announced that Zonds 4, 5 and 6 were aimed at perfecting a manned spacecraft to go around the Moon, and that further unmanned tests would be required before men could be flown.

Zond 7

Zond 7 was launched on 7 August 1969, and was basically a repeat of the Zond 6 mission, although it took colour as well as black and white photographs of the Moon and the Earth.

It was launched a fortnight after the Americans successfully landed on the Moon.

Zond 8

Zond 8, launched on 20 October 1970, was the last in the Zond series. It was basically a repeat of the Zond 7 mission, but used a very different approach to Earth for re-entry, coming in over the North Pole instead of the South. This enabled Soviet ground stations to control the flight during most of the re-entry. The spacecraft again splashed down at night in the Indian Ocean.

Cosmos 379, 382, 398 and 434

Four Cosmos flights launched in 1970-71 are thought to have been connected with the Soviet lunar programme, although they were not given the Zond designation.

Cosmos 379, 398 and 434 all had orbital parameters and manoeuvres which were very similar to each other. Their initial orbits, signal formats and the frequencies used were similar to a Soyuz mission, but the apogees were raised to as much as 8721 miles (14,035 km). The Soyuz on the other hand had been described as having a maximum altitude range of 808 miles (1300 km).

Western calculations have shown that the Cosmos 379 manouevres approximated to what might be expected for lunar orbit insertion and trans-Earth injection.

The Soviets were forced to reveal the nature of Cosmos 434 in August 1981, when it was feared that it had re-entered over Australia. To allay public fears that it might pose a radioactive hazard in the same way as Cosmos 954, an official statement was issued by the Soviet Foreign Ministry. This revealed that Cosmos 434 had been a test of a 'lunar cabin.' Whether this cabin was a lander or orbiter was unclear. As Cosmos 379 and 398 were so similar to Cosmos 434, there appears to be little doubt that they were also part of the Soviet manned lunar programme.

Cosmos 382 was very different to the other three Cosmos flights in that its perigee was raised, rather than its apogee. It also made a very substantial plane change in its final manoeuvre.

Some Western observers have concluded that all four Cosmos flights were a test of the Soviet equivalent of the American Apollo SPS (Service Propulsion System) engine. It has also been speculated that Cosmos 382 was actually a test of a lunar orbiter and lander.

Mission Profile

This section will attempt to show possible lunar landing mission profiles, but in view of the lack of official information, must remain highly speculative.

Whereas the American Apollo programme launched the crew and all mission spacecraft on one rocket – the Saturn V – it appears that the Soviets intended to use two launchers. The G-vehicle would have launched the lunar spacecraft, and a Proton D booster would have carried the fully loaded manned craft that would return the crew to Earth. It is believed that the normal Soyuz A-2 booster would not have had the capacity for such a mission. The manned spacecraft would be launched first, and following a successful G-vehicle launch, would then carry out a rendezvous and docking with it in Earth orbit.

The rendezvous and docking may have been tested on the two occasions where pairs of Cosmos satellites were docked in orbit (Cosmos 186/188 and Cosmos 212/213). A similar manned Soyuz docking was accomplished in 1969 (Soyuz 4/5). On each mission the active vehicle was launched first, followed by the passive target. The same sequence would have been required if a manned spacecraft had to dock with a G-vehicle lunar landing complex before the giant booster's upper stage fuel became unusable.

It would appear that during all but the last of the G test launches there was a simultaneous Proton booster countdown. However, only during the 3 and 4 July 1969 double count was there any actual manned crew involved. (see Crewing Assignments below).

It is thought that the crew would have been launched into Earth orbit on a heavy Zond i.e. a modified Soyuz spacecraft with a full propellant load in its base torus tank. As well as the heavy Zond, the Proton vehicle would have been able to carry another module for such functions as in-orbit propulsion, life support or overall mission support. This would have been desirable as the basic heavy Zond had limited provisions.

Following Earth orbit rendezvous, the crew would have jettisoned any extra modules before docking with the G manned lunar complex. It is believed that the G-vehicle would have carried the Soviet Lunar Module, the lunar insertion stage, and a mission support module.

The Mission Support Module was intended to provide support functions during the trans-lunar and lunar-orbit phases of the mission, and the Lunar Braking Module would have been used for lunar orbit insertion and orbit trim. The Soviet Lunar Module is believed to have been designed to carry two cosmonauts to the lunar surface.

Upon completion of the lunar mission, the heavy Zond propulsion system would have been used to perform the trans-Earth injection manoeuvre.

Crewing Assignments

In 1963, Yuri Gagarin became Commander of the Cosmonaut Detachment, and was also made responsible for all projects connected with lunar exploration. Because of this, and because he insisted that he would make a second, possibly third spaceflight, many people believed that he would lead the Soviet lunar landing team. Apparent confirmation came from Valentina Tereshkova when, during a visit to Havana, she told the Cubans 'The Moon team has already been picked. Major Gagarin is head of it, and I am on it.'

However, it now appears that Gagarin and Tereshkova were never really part of the lunar programme. In 1981, Vitaly Sevastyanov confirmed that there had been a Soviet lunar programme, and he identified the cosmonauts involved. The team was formed in the middle of 1967, and as well as training on helicopters, they practiced splashdown procedures in the Black Sea. The group consisted of Artyukhin, Bykovsky, Dobrovolsky, Grechko, Klimuk, Leonov, Makarov, Popovich, Rukavishnikov and Sevastyanov, the latter being the training manager. Sevastyanov was replaced in that role by Pavel Belyayev in late 1967 in order that he could join the main Soyuz programme.

Belyayev started training for a Zond circumlunar flight in late 1967, and it appears that he should have been launched in December 1968. However the flight was apparently cancelled at the last moment. His back-up is thought to have been Valery Bykovsky, and both men were present at Baikonur in December 1968. (Other Western reports have stated that Leonov was Belyayev's back-up).

Soviet tracking and recovery ships were deployed in the first week of July 1969, indicating an imminent manned mission. It is thought that a high Earth orbital mission was planned, and it is believed that a heavy Zond-Proton type double countdown was conducted on 3 and 4 July 1969.

According to C.P. Vick, (see below), during the countdown, a cosmonaut crew was present at Baikonur, and there are indications that this crew was composed of Anatoly Filipchenko, Valery Kubasov and a third man, possibly Georgi Shonin. They were apparently involved in the heavy Zond-Proton part of the mission, and their launch was cancelled when the G-vehicle failed soon after lift-off, collapsing back onto the launch platform.

Private conversations conducted with participants in the Soviet Lunar programme by C.P. Vick (see Further Reading below) confirmed the Belyayev/Bykovsky flight and revealed some of the flight-engineer assignments:

Oleg Makarov was the overall Chief Soviet Lunar Module (SLM) Engineer; Georgi Grechko was Chief SLM electronics instrumentation engineer; and Vitaly Sevastyanov was one of the prime crewmen for the first outbound manned Lunar mission.

The flight-engineer assignments were:

Kubasov — Earthorbital (SLM) test flight;

Sevastyanov — Lunar Orbit (SLM) test flight;

Makarov — 1st Lunar Landing;

Grechko — 2nd Lunar Landing;

Rukavishnikov — 3rd Lunar Landing.

The Kubasov and Sevastyanov assignments are somewhat puzzling. By his own admission, Sevastyanov was switched out of the lunar programme into the Soyuz programme, and Kubasov was never a member of the lunar training group. The above assignments must therefore be viewed as somewhat speculative.

The commanders were presumably chosen from Belyayev, Bykovsky, Dobrovolsky, Klimuk, Leonov and Popovich i.e. the remaining men from the 1967 lunar landing team. Artyukhin, a non-pilot, would have been a reserve flight-engineer.

Further reading

As stated in the introduction, it is beyond the scope of this book to give a detailed analysis of the Zond programme. Those readers wishing to study the subject in more detail are advised to read the following:

C.P. Vick 'The Soviet G-1-e Manned Lunar Landing programme Booster' JBIS January 1985.

D.R. Woods 'A Review of the Soviet Lunar Exploration programme' SPACEFLIGHT July-August 1976.

J.E. Oberg 'Russia Meant to Win the Moon Race' SPACEFLIGHT April 1975.

1.18: The Yuri Gagarin Cosmonaut Training Centre

Introduction

The Yuri Gagarin Cosmonaut Training Centre is the principal test and training facility for Soviet cosmonauts. This section examines its history and capabilities, its hierarchy, the training of Soviet cosmonauts, and finally, the training of the Intercosmos crews, which gave new insights into Soviet training practices and procedures.

Background

The Cosmonaut Training Centre is located about 25 miles (40km) north-east of Moscow, in Zvezdny Gorodok (Star Village, or Star Town.) The Soviets refer to it as TPSK (or TSPK) – Tsentr Podgatovka Kosmonavtov – the Training Centre for Preparing Cosmonauts.

It was completed in September 1960, and originally consisted of just two buildings which were used for training the first cosmonauts. The reason for establishing Star Town was two-fold: firstly to have, under one roof, everything that was needed to train cosmonauts (aircraft, centrifuge, hydrotank, simulators and other equipment) and secondly, to have a community in which the cosmonauts, engineers, scientists and doctors could live. (Those workers who commute into Star Town do so in cars with special numberplates.) Star Town was born, according to Georgi Beregovoi, on 11 January 1960, when the first cosmonauts began training with the minimum of facilities: a rotating chair, a chair that could spin 360° and a medical laboratory.

The Cosmonaut Training Centre was renamed 'The Yuri Gagarin Cosmonaut Training Centre' in 1968 following the death of Yuri Gagarin. In 1971, it was awarded the Order of Lenin for 'its great services in the preparation of crews for spaceflights, its participation in the mastery of space, and in connection with the 10th anniversary of the world's first manned spaceflight.'

Facilities at the centre include classrooms, laboratories, scale models of spacecraft and space-stations, a sports training complex, centrifuges, thermal chambers and other equipment to simulate conditions in space. Among the latter is a pool, or hydrotank, 25 yards (23m) in diameter and 13 yards (12m) deep, containing models of a Soyuz craft and a Salyut station. In this neutral buoyancy facility, cosmonauts wearing spacesuits practice extravehicular activity.

In 1973, American officials visited Star Town as part of the preparations for ASTP, and reported that a significant amount of new construction was underway, including a new training building especially for ASTP. The four-storey building was to include classrooms, lecture halls, and display rooms for spacecraft sub-systems. In addition, the Soviets were building a new hostel and dispensary for the US team, as well as other buildings to house simulators and a new and very large centrifuge.

In 1985, it was reported that a new simulator had been built to train cosmonauts for assembly work in space. Participants in future space-station missions are expected to assemble various structures in orbit that will first be delivered by unmanned spacecraft in parts.

Star Town provides good living accomodation for the cosmonauts and their families, and associated scientists and technicians. It is under the administration of the Soviet Academy of Sciences, although the Soviet Air Force is responsible for cosmonaut training.

Crews usually leave Star Town for Baikonur 21 days before a flight.

The Hierarchy of the Yuri Gagarin Cosmonaut Training Centre

The cosmonauts at the Training Centre include cosmonauts preparing for future space missions as well as cosmonauts from previous missions serving in management and instructor roles. General Georgi Beregovoi is the Head of the Cosmonaut Training Centre, and the other known administrative positions are as follows:-

First Deputy Head	Andrian Nikolayev
Deputy Head	Pavel Popovich
Deputy Head and Head of the International Detachment	Alexei Leonov
Deputy Head for Political Affairs	Pyotr Klimuk
Commander of the Cosmonaut Detachment	Boris Volynov
Director of Cosmonaut Training	Vladimir Shatalov
Head of a Training Department	Viktor Gorbatko
Head of a Training Department	Georgi Shonin
Training Managers	Yuri Glazkov, Alexei Gubarev and Vasily Lazarev.

Soviet Cosmonaut Training

The various training protocols for Soviet cosmonauts still demand rigorous physiological and psychological stress training, in those areas of spaceflight considered critical for the successful accomplishment of the mission. In particular, based on actual experience gained in prior manned spaceflights, training emphasis is placed in areas considered to be the most troublesome to the cosmonaut. The overriding problem influencing all training efforts is to moderate, if possible, the effects of weightlessness on physiological well being. Since the inception of manned spaceflight a rather specific series of problem areas have been defined. These are all related in some degree or other to the weightless state and consist of motion sickness, circulatory deconditioning, red blood cell mass loss and bone demineralisation. In addition, isolation may be a serious psychological problem.

As part of their preparations, cosmonauts undergo a variety of conditioning programmes which include vestibular, visual, acceleration, weightlessness, survival and psychological training.

One of the major problems encountered by Soviet cosmonauts during actual spaceflights has been vestibular disorders or space motion sickness. This disorder is still not fully understood, and was first encountered in 1961 when Gherman Titov reported periods of disorientation and nausea during the Vostok 2 mission.

Since then, the Soviets have devoted extensive pre-flight training time and have utilised exposure to pre-flight vestibular stress in selecting cosmonauts less prone to motion sickness. Pre-flight training involves using both passive and active exposure to increase vestibular tolerance and is tailored to the individual cosmonaut. Active conditioning is a process where the individual cosmonaut participates in specific strenuous physical exercises as well as using gymnastic equipment. Passive conditioning includes the cosmonaut only in the sense that he or she is strapped into or is otherwise an integral component of a piece of equipment that is mechanically activated and either rotates about several axes or in some manner stresses the vestibular and optical system. The Soviets report that this conditioning reduces gastrointestinal space motion sickness.

All aspects of the training programme are directed at adapting the individual cosmonaut, as much as possible, to subsequent space stress condition by repeated exposure, with gradually increased intensity, to a variety of these external stimuli. In all instances, individual physical, pyschological and physiological profiles are taken into consideration in order to arrive at exposure and optimal load conditions appropriate for each cosmonaut.

In order to provide the individual cosmonaut with the capacity to adapt to transverse acceleration during spaceflight, particularly during initial lift-off or other phases of rapid acceleration, extensive familiarization runs are made in centrifuges. During such an exposure, individual tolerance levels are significantly increased to transverse acceleration stress, as is the capacity of the cosmonaut to develop the ability to respond to such a stress and thereby reduce its adverse impact. This is augumented by means of cosmonaut leg muscle tensing, and abdominal and thoracic breathing. Centrifugal forces are increased gradually over periodic exposures during a 2 month interval. Cosmonauts have been exposed to transverse G forces of up to 10 G. Soviet scientists have demonstrated that gradually increasing G-force exposure is effective in enhancing tolerance levels. They also suggest that the best responses have been attained when G-forces are gradually increased, whilst at the same time the intervals between G-force exposure are also gradually increased.

Although weightlessness for any length of time is not readily reproduced on Earth, it can be replicated for short intervals by aircraft flights along Keplarian (parabolic) trajectories. The use of such flights for conditioning purposes have shown that human response is individualistic, but that repeated exposure of individuals does permit a degree of adaptation even for individuals who initially have a low tolerance.

Soviet investigators have demonstrated that individuals who are experienced flight personnel in general are more tolerant to short term weightlessness exposure and adapt more readily during repeated exposure than do non-flight personnel. These types of studies, as well as more than 25 years experience, have led some Soviet space scientists to conclude that short term exposure of spaceflight personnel to parabolic flight trajectories are beneficial both for training purposes as well as for determining predisposition to vestibular disorders.

An integral component of acceleration and vestibular training is visual perception training. The eye is thought to be a contributing sensory factor associated with space motion sickness.

Physical conditioning demands have been lowered and the age of acceptable candidates raised. In particular, such requirements as spatial orientation and visual perception training was lowered for science crewmembers, since the latter would not have a primary requirement to pilot the spacecraft. However, tolerance to G forces, and vestibular stability are still a major requirement for all cosmonauts.

Training procedures aid the cosmonaut in learning how to operate the spacecraft prior to actually going into space. This assures that a variety of experiences and sensations are simulated and recognised during the training phase of the programme, thereby allowing the cosmonaut to respond in emergency situations in a similar manner to that while undergoing training. Weightlessness was first simulated by using a swimming pool, and a special hydrotank is now used. This facility is monitored visually via television and other electronic means during the time the cosmonauts are submerged in a pool in their Salyut and/or Soyuz simulator 13 yards (12m) below the surface. In addition, the Training Centre now utilises an IL-76 aircraft for parabolic flights to enable cosmonauts to experience brief exposure to zero gravity.

Other facilities are available to acquaint the cosmonauts with expected flight conditions, and these include a space planetarium to familiarize them with celestial bodies used for space navigation. This facility also permits them to become familiar with the use of celestial bodies that will be studied as part of the scientific programmes. Since the amount of research carried out in space increases continually and since the same individuals may be required to perform a multitude of such studies, it is essential that they are adequately trained on Earth. The Soviets also seem to incorporate into cosmonaut training an understanding and appreciation of the Marxist-Leninist principles, which provide not only the technical rationale for the mission, but also social and political reasons.

As space missions have increased in duration, the process of selection and training of cosmonauts has evolved into training of both the body and mind. The crew of a spacecraft should consist of individuals who are compatible.

Despite the selection and training process, psychological tensions between crewmembers do develop. Therefore appreciable training time is devoted to the behavioral areas, including response to emergency conditions potentially encountered in space. Since individuals are confined for long intervals in limited quarters under conditions requiring special safety and life support systems, a detailed psychoanalysis of the individuals who would be most compatible in such environments is necessary. To accomplish this type of psychological selection and training, individuals are frequently subjected to multiple and sequential stresses such as heat, humidity, cold, and isolation.

Factors that determine group psychology and compatibility are analysed. These include intellectual abilities, and likes and dislikes. It is very important that members of a group have a positive attitude toward each other and are willing to co-operate and conform in order to accomplish their mission.

Newer methods of evaluating stress responses are being investigated. These include brain scans, voice stress tests, and changes in circulatory endocrine levels. As techniques for quantifying physiological and psychological stress responses improve, the selection and training of cosmonauts, as well as modification of the space capsule environment, may be become more of a quantitative science.

It has become apparent from Salyut missions that psychological support for the crew during spaceflights of long duration is highly effective in maintaining motivation and emotional stability. The psychological support programme is based on a series of informational exchanges and conversations between the cosmonauts and ground control, family members, scientific and technical specialists, as well as entertainers and politicians. All of these support activities are, of course, also utilised during psychological stress training practiced on the ground which attempt to emotionally prepare the cosmonaut for the unexpected.

Soviet training programmes prepare the individual for stresses other than crisis such as isolation, boredom and confinement. The Soviet philosophy in this area of training is based on the supposition that equipment reliability is not endless. The ultimate success of a mission may very well depend on the psychological fine tuning of the cosmonaut. Therefore, they incorporate the various forms of stress training such as underwater tank diving, centrifugation, altitude chamber rides, isolation, parachuting, and survival training in their overall emotional and physical conditioning programmes. The cosmonauts' parachute training involves at least 100 jumps of increasing difficulty. Trainees initially must count to 20 seconds prior to opening their parachutes. This is followed by requiring them to go through a check list during free fall. Later, they must await ground control approval prior to opening their parachutes. Survival tests include depositing of the cosmonauts and their space capsule trainer in remote, environmentally hostile areas where they are left to their own devices to survive and return to safer terrain. In addition, splashdown procedures are practiced in the Black Sea.

As previously mentioned, the Soviets have placed emphasis on formal physchological training. This is apparent by their incorporation of psychological monitoring equipment onboard spacecraft, including electronic equipment which monitors tension and mood changes by voice harmonics analysis and changes in facial gestures. The cosmonauts are kept under constant observation by doctors and psychologists using onboard TV cameras.

Training of Intercosmos crews

The Intercosmos cosmonauts trained intensively for 1½ to 2 years at the Cosmonaut Training Centre. The first stage included Russian language lessons, and theoretical studies combined with flight experience in jet planes, physical exercises, simulation of zero-G flights, practicing elementary procedures of splashdown, and retrieval by helicopters from thick forests, mountains and other difficult terrain. The men underwent survival training which included simulating an off-target landing in winter tundra or waterless desert, during which they had to survive for a few days using the small emergency rations of water, clothes and food available onboard the Soyuz spacecraft.

The cosmonauts also carried out theoretical studies, absorbing the fundamental dynamics of spaceflight, mathematics and ballistic calculations, the fundamental design of space vehicles, astronomy and astro-navigation, and computers.

Their working day exceeded 12 hours, following which it was usual to study at home for a few hours and prepare for the next working day. In addition, there were obligatory clinical and physiological check-ups every three months, so good health was essential.

According to Alexei Leonov, Soviet spaceflight experience has shown that in zero-G man needs stamina rather than strength, as well as balanced psychological reactions, instantaneous reactions to a changing situation and considerable reserves of the cardiac and vascular system. Therefore sports are an important part of the general training of cosmonauts.

The initial period of training for the first Intercosmos cosmonauts ended with examinations in more than a dozen special scientific subjects.

The second stage consisted in mastering the Soyuz spacecraft and the specific programme for a visit to the Salyut space-station. During this period, the training was conducted at crew strength, and the future research-cosmonauts were acquainted with their commanders, all of whom were experienced cosmonauts.

The two French cosmonauts, Jean-Loup Chrêtien and Patrick Baudry, found the Soviet training philosophy significantly different from the flight training for the US Shuttle. One of the most noticeable differences was the absence of basic documentation, such as systems manuals. Instead of being supplied with such manuals, the cosmonauts had to compile notes during their training and refer back to them during the mission.

The training courses they attended were based on a repetitive learning process, and consisted of lectures by specialists on specific subjects. The lectures were repeated from time to time throughout the training programme.

'The instructor for the Soyuz T computer system, for example, would give a lecture on the computer from A to Z, and then come back and give the same lecture,' one French programme official said. 'This is their philosophy – learning through repetition.'

1.19 The Baikonur Cosmodrome (Tyuratam)

Introduction

All Soviet manned spaceflights are launched from the Baikonur Cosmodrome, which is the Soviet Union's largest launch site. From the inception of the Soviet space programme, controversy has raged over the choice of name for the site, and the deliberate Soviet attempts to mislead the West over its precise location. This section examines the history of the Cosmodrome, its facilities, and location.

Background

The largest and most versatile of the Soviet launch sites is located near the rail stop village of Tyuratam in Kazakhstan at about 45.6°N. latitude, 63.4°E. longitude. The Soviets refer to it as the Baikonur Cosmodrome, although it is about 230 miles (370 km) southwest of the station stop of that name on quite a different railway line. It is thought that by giving contradictory information about the Cosmodrome in the early days of their space programme, the Soviets hoped to maintain some element of doubt in the West, since the town of Baikonur is on the correct ground trace of the early Soviet flights which were at

65° inclination. However, even now, the Soviets pinpoint the launch pad for manned flights as being at 47.3° N. and 65.5° E. which is patently false in the light of conclusive public evidence of initial revolution ground traces and known launch times. During the ASTP project, even NASA gave the launch site as being at 47.8° N. and 66° E., which was in direct contradiction to NASA Landsat photographs and the visits and descriptions supplied by NASA visitors to the launch site.

Low resolution Landsat 1 photographs available to the public in the 1970's showed that the Tyuratam area was covered with roads, railway tracks, and other signs of human activities including almost certainly assembly buildings and launch pads which spread over a distance of about 84 × 56 miles (135 × 90 km) or more. More recent Landsat imagery suggests that additional development at either extremity has increased the East-West dimension to nearer 93 miles (150 km.) With launch pads for many different launch vehicles scattered over the area, it is not possible to speak of a single closely defined latitude and longitude as defining the site, or to know what all the launch facilities look like.

The Landsat pictures of the Cosmodrome confirm that it is situated in open steppe country, relatively flat and only slightly rolling.

Baikonur was the site from which the first Soviet ICBM's, all the early Sputniks, all manned flights, all lunar and planetary flights, and the earlier communications satellites were launched. The same launch pad used by Sputnik 1 has been used for most manned launches since.

The first Westerner to visit the launch site was President De Gaulle of France, in June 1966. In 1970, President Pompidou also witnessed a military satellite launch from the Cosmodrome. And in the early 1970's, three parties of American astronauts and technicians were permitted to visit the Cosmodrome during the preparations for ASTP. However, they were flown in at night, put up in a hotel, driven to the launch pad, and then were returned to their hotel for another night time flight out. Questions had to be limited to the launching pad, rocket, and spacecraft they would be on. Requests to see the underground blockhouse were turned down. Moreover, American newsmen were not permitted to witness the lift-off of Soyuz 19.

At the time of ASTP, the Soviet Union stressed their intention to have a back-up Soyuz spacecraft, launcher and crew ready and waiting on a separate pad to ensure their ability to honour their committment to launch, should a major problem prevent the launching of the prime crew. The distance between the two pads was said to be about 19 miles (30km.)

In 1978, it was reported that US reconnaissance spacecraft had detected large new runway construction at Baikonur, and it was suggested that this was being built for the Soviet space shuttle programme.

Leninsk

As already stated, the Baikonur Cosmodrome is located near the rail stop village of Tyuratam. This has now been completely overwhelmed by the adjacent city of Leninsk, which has a population of about 50,000 people. This city has not been shown in public Soviet atlases, and seems to owe its existence to the space programme.

It is clearly seen in photographs of the area taken by Landsat north of Tyuratam. The town of Baikonur is situated more than 150 miles (240 km) to the north-east. This has prompted many Western observers to ask whether the Cosmodrome should now be properly referred to as the 'Leninsk Cosmodrome' instead of 'Baikonur' or 'Tyuratam.'

Leninsk is said to be about 1300 miles (2090 km) south-east of Moscow on the main Moscow-Tashkent railway line. The area has been described as rolling but mostly flat country, and the city is said to be interlaced by a complex irrigation system and dotted with tall trees. The climate is very extreme in summer and winter, and 40° C of frost with violent snow-storms is not unusual. It is said to be 20 miles (32km) from Leninsk to the launch pad used for Sputnik 1, Vostok 1, and the Soyuz 19 ASTP spacecraft.

There is a cosmonaut training complex at Baikonur, presumably in or near Leninsk. The facilities include classrooms for technical and scientific training, laboratories to prepare the cosmonauts for flight, a sports complex or stadium, a 164 ft. (50 m) swimming pool, and a medical complex.

In addition, Leninsk has many schools, several 'palaces of culture,' cinemas, and cafés and restaurants. The local TV centre broadcasts its own programmes.

The American visitors during ASTP found the accomodations provided at Leninsk to be equal or superior to those provided at the Kennedy Space Centre. The cosmonauts themselves have comfortable accomodation in cottages and there is a hotel complex with all the necessary equipment for pre-launch medical checks. This facility, together with the sports stadium and swimming pool, lies on the outskirts of Leninsk, not far from the main gateway of the town where there is a blue sign with a white arrow pointing 'To Baikonur'. The highway, which passes near the Tyuratam railway station, is lost amidst low hills and then radiates towards numerous launch pads.

According to one Western report, the Soviets refuse to recognize, and actually detest, the Tyuratam designation given to the Cosmodrome by the US Government. According to the report, the Cosmodrome was named by a CIA official using a German war-time map, simply because it was the most accurate then available. Soviet annoyance could have been intensified because Tyuratam apparently means 'arrow burial ground' – which could hardly be a more inappropriate name for a rocket test centre.

Spacecraft Assembly and Test

The assembly-test building used for the Soyuz is called the MIK (Montazhno Ispytatel'nyy Korpus). The Soyuz is first given a full check-out in the MIK, and then again on the pad. In the MIK, the separate modules are tested in vacuum chambers, including the firing of manoeuvering engines. After the individual modules are tested, they are assembled to create the whole vehicle and returned to the vacuum chamber for further check-out. Then they are placed in an anechoic chamber to test the radio compatibility of the assembled ship with its communications system.

Check-out of the spacecraft is done in the vertical position, whereas the mating of the spacecraft and the launch vehicle is done horizontally. The launch vehicle is then carried horizontally on railway flatcars to the launch pad, where it is tilted up to sit on a stand over a large flame detector pit. The MIK is located about 1 mile (1.6km) from the launch pad, and the two are joined by standard Soviet 5 foot gauge railway track.

Two small 'memorial houses' not far from the MIK have been preserved because of their historic association with Sergei Korolyov. Yuri Gagarin spent the night before starting his historic space journey in one of them, and Korolyov stayed in the other during his frequent visits to the Cosmodrome.

Cosmonaut launch preparations

Crews leave for Baikonur 21 days before a flight, and are exposed to a minimum of contact with other people. They do light simulator work, exercise, and study documents and schedules, although they do no flying. They stay in the Cosmonauts Hotel, and just before their flight, the prime crew and the back-up crew plant a tree in a special park.

Once suited up for flight, the cosmonauts ride out to the launch pad in a well-equipped, air-conditioned bus, much in the manner that NASA astronauts are transported.

Following their return to Earth, the crew remains at the Cosmodrome for about 15 days before returning to Star Town.

1.20: Callsigns

Introduction

Each Soviet cosmonaut has his own callsign which is used during communication sessions. However, during missions involving more than one crewmember, the remainder of the crew adopt the same callsign as their commander. For example, Filipchenko's callsign is Buran (Snowstorm), and therefore, on Soyuz 16, his flight-engineer – Rukavishnikov – was given the callsign Buran-2.

If the other members of the crew subsequently make another spaceflight, but this time as commander, they then use their own callsign. Again taking the example of Rukavishnikov, when he was given command of the Soyuz 33 mission he dropped the Buran-2 callsign in favour of his own i.e. Saturny (Saturn). Naturally, his companion – Ivanov – was given the callsign Saturny-2.

The normal method of referring to a crew is by calling them, for example, 'the Buran crew', rather than speaking of 'Buran-1 and Buran-2'.

Callsigns are also used during training sessions, adopting the same system as for actual missions. For example, during training for the joint Soviet – East German Soyuz 31 flight, the back-up crew of Gorbatko and Köllner used the callsigns Terek 1 and Terek 2 respectively, as they would have done had they actually flown.

There is apparently no fixed system of assigning callsigns, although certain patterns can be established. All the Vostok callsigns, with the exception of Gagarin, were names of birds, and both Voskhod callsigns were precious stones. However, once into the Soyuz programme, the pattern becomes less clear. Broadly speaking, the Soyuz callsigns can be broken down into a mixture of scientific and astronomical terms, together with geographical features of the Soviet Union.

Whilst all the cosmonauts have callsigns, these are only disclosed when the cosmonaut in question commands a mission. Therefore, there are many men for whom the **individual** callsign is unknown.

N.B. Flight Control Centre always uses the same callsign – 'Zarya' (Dawn).

Mission Callsigns

Mission	Callsign	Commander
Vostok 1	Kedr	Gagarin
Vostok 2	Oryel	Titov
Vostok 3	Sokol	Nikolayev
Vostok 4	Berkut	Popovich
Vostok 5	Yastreb	Bykovsky
Vostok 6	Chayka	Tereshkova
Voskhod 1	Rubin	Komarov
Voskhod 2	Almaz	Belyayev
Soyuz 1	Rubin	Komarov
Soyuz 2	Unmanned	Unmanned
Soyuz 3	Argon	Beregovoi
Soyuz 4	Amur	Shatalov
Soyuz 5	Baikal	Volynov
Soyuz 6	Antey	Shonin
Soyuz 7	Buran	Filipchenko
Soyuz 8	Granit	Shatalov
Soyuz 9	Sokol	Nikolayev
Soyuz 10	Granit	Shatalov
Soyuz 11	Yantar	Dobrovolsky
Soyuz 12	Ural	Lazarev
Soyuz 13	Kavkas	Klimuk
Soyuz 14	Berkut	Popovich
Soyuz 15	Dunay	Sarafanov
Soyuz 16	Buran	Filipchenko
Soyuz 17	Zenit	Gubarev

April 5 anomaly	Ural	Lazarev
Soyuz 18	Kavkas	Klimuk
Soyuz 19	Soyuz	Leonov
Soyuz 20	Unmanned	Unmanned
Soyuz 21	Baikal	Volynov
Soyuz 22	Yastreb	Bykovsky
Soyuz 23	Rodon	Zudov
Soyuz 24	Terek	Gorbatko
Soyuz 25	Fotony	Kovalyonok
Soyuz 26	Taimyr	Romanenko
Soyuz 27	Pamir	Dzhanibekov
Soyuz 28	Zenit	Gubarev
Soyuz 29	Fotony	Kovalyonok
Soyuz 30	Kavkas	Klimuk
Soyuz 31	Yastreb	Bykovsky
Soyuz 32	Protony	Lyakhov
Soyuz 33	Saturny	Rukavishnikov
Soyuz 34	Unmanned	Unmanned
Soyuz T	Unmanned	Unmanned
Soyuz 35	Dnepry	Popov
Soyuz 36	Orion	Kubasov
Soyuz T-2	Yupiter	Malyshev
Soyuz 37	Terek	Gorbatko
Soyuz 38	Taimyr	Romanenko
Soyuz T-3	Mayak	Kizim
Soyuz T-4	Fotony	Kovalyonok
Soyuz 39	Pamir	Dzhanibekov
Soyuz 40	Dnepry	Popov

Mission	Callsign	Commander
Soyuz T-5	Elbrusy	Berezovoi
Soyuz T-6	Pamir	Dzhanibekov
Soyuz T-7	Dnepry	Popov
Soyuz T-8	Okeany	Titov
Soyuz T-9	Protony	Lyakhov
Launch abort	Okeany	Titov
Soyuz T-10	Mayak	Kizim
Soyuz T-11	Yupiter	Malyshev
Soyuz T-12	Pamir	Dzhanibekov
Soyuz T-13	Pamir	Dzhanibekov
Soyuz T-14	Cheget	Vasyutin

Cosmonaut Callsigns

Cosmonaut	Callsign	Meaning	Notes
Aksyonov	Unknown		
Alexandrov	Unknown		
Andreyev	Unknown		
Anokhin	Unknown		
Artyukhin	Unknown		
Atkov	Unknown		
Belyayev	Almaz	Diamond	Precious stone.
Beregovoi	Argon	Argon	An inert gas.
Berezovoi	Elbrusy	Elbrus	One of the Caucasus mountains in the south of the Soviet Union, it has two extinct volcanic peaks and is the highest mountain in Europe.

Cosmonaut	Callsign	Meaning	Notes
Bykovsky	Yastreb	Hawk	self-explanatory
Demin	Unknown		
Dobrovolsky	Yantar	Amber	Yellow fossil resin.
Dzhanibekov	Pamir	Pamirs	A mountain range in Central Asia.
Feoktistov	Unknown		
Filipchenko	Buran	Snowstorm	self-explanatory
Gagarin	Kedr	Cedar	1
Glazkov	Unknown		
Gorbatko	Terek	Terek	A river on the Caucasus.
Grechko	Unknown		
Gubarev	Zenit	Zenith	The highest point.
Illarionov	Unknown		
Ivanchenkov	Unknown		
Kizim	Mayak	Beacon	self-explanatory
Khrunov	Unknown		
Klimuk	Kavkas	Caucasus	Soviet mountain range.
Komarov	Rubin	Ruby	Precious stone.
Kovalyonok	Fotony	Photon	Unit of energy.
Kubasov	Orion	Orion	The Constellation.
Lazarev	Ural	Urals	Soviet mountain range.
Lebedev	Unknown		
Leonov	Soyuz	Union	2
Lyakhov	Protony	Proton	Positive particle in the nucleus of an atom.
Makarov	Unknown		

Cosmonaut	**Callsign**	**Meaning**	**Notes**
Malyshev	Yupiter	Jupiter	The planet.
Nikolayev	Sokol	Falcon	self-explanatory
Patsayev	Unknown		
Popov	Dnepry	Dnieper	A Soviet river in the Ukraine.
Popovich	Berkut	Golden Eagle	self-explanatory
Romanenko	Taimyr	Taimyr	A large Soviet peninsula.
Rozhdestvensky	Unknown		
Rukavishnikov	Saturny	Saturn	The planet.
Ryumin	Unknown		
Sarafanov	Dunay	Danube	The European river.
Savinykh	Unknown		
Savitskaya	Unknown		
Serebrov	Unknown		
Sevastyanov	Unknown		
Shatalov 1	Amur	Amur	A Soviet river[3]
Shatalov 2	Granit	Granite	[3]
Solovyov	Unknown		
Shonin	Antey	Antaeus	Mythical giant[4]
Strekalov	Unknown		
Tereshkova	Chayka	Seagull	self-explanatory
Titov, G	Oryel	Eagle	self-explanatory
Titov, V	Okeany	Ocean	self-explanatory
Vasyutin	Cheget	Cheget	A mountain peak in the Caucasus. (Alternative spelling is Tcheget.)
Volk	Unknown		

Volkov, A	Unknown		
Volkov, V	Unknown		
Volynov	Baikal	Baikal	A Soviet lake[3]
Yegorov	Unknown		
Yeliseyev	Unknown		
Zholobov	Unknown		
Zudov	Rodon	Radon	Radioactive element.

Notes

1: All the Vostok callsigns but for Gagarin's were names of birds – no explanation has been given as to why he was the exception. However, in some books he is reported as using the callsign 'Lastochka' or 'Swallow', which would fit the pattern of the other Vostok pilots. All official Soviet histories however, refer to the 'Kedr' callsign.

The origin of the 'Lastochka' story is unclear, but it has been suggested that the blame lies with Kruschev, who was quoted in April 1961 as calling Gagarin 'Space Swallow.' It would seem likely that this term was erroneously believed to be Gagarin's callsign by later space historians, especially as it fitted the known pattern.

2: It is unlikely that 'Soyuz' was Leonov's real callsign. During the Apollo-Soyuz Test Project, the US spacecraft was given the callsign 'Apollo.' What could be simpler than to call the Soviet spacecraft 'Soyuz'? During pre-flight preparations, Leonov stated that once the joint operations had been completed, he would use the callsign 'Almaz' - the callsign originally used by Pavel Belyayev, his commander onboard Voskhod 2. However, it is not known if he went ahead with this plan.

A clue to Leonov's real callsign is contained in his book 'The Sun's Wind', where he states that all the cosmonauts had nicknames or callsigns, and that his was 'Blond'. It appears however, that this was simply a nickname, rather than his callsign, as it does not fit the general pattern of callsigns.

3: The use of the callsigns Amur and Baikal had a special significance on the Soyuz 4/5 joint flight – the first Soviet manned docking mission.

Amur is the name of a river running along the boundary of Manchuria and the Soviet Union. Baikal is a lake in Southern Siberia – the largest freshwater lake in Asia.

For many years, the Soviet Union had been constructing a gigantic railway line joining the two, known as the Baikal-Amur Magistral, or BAM, project. The use of the callsigns Amur and Baikal therefore was a symbolic gesture, a salute to the workers on the railway link-up from the cosmonauts involved in the space link-up.

Interestingly, Shatalov, or 'Amur', subsequently dropped this callsign, and chose 'Granit' for his next two flights. He thereby became the only mission commander to have used two separate callsigns.

4: Antaeus is a character from Greek mythology. He was an African giant who was invincible as long as he touched the ground. Hercules therefore lifted him into the air before crushing him to death.

This appears an extremely odd choice of callsign, as Shonin is seemingly saying that he is, like Antaeus, especially vulnerable when he leaves the ground!

1.21: CapCom Assignments

Introduction

The CapCom (or Capsule Communicator) is the term given to the cosmonaut who is present at the Flight Control Centre during a mission and is responsible for voice communications with the crew. The assignment is usually given to a cosmonaut familiar with the mission, often a member of the back-up or support crew, and it is therefore a valuable training role. This section gives background information about the FCC and examines the known CapCom assignments.

The tables which follow are incomplete as the Soviets do not always disclose the CapCom assignments for a mission. Those entries shown in *italics* are where a cosmonaut has been reported as being 'involved with the flight control' of a mission i.e the exact role has not been disclosed and a CapCom assignment has been assumed.

Flight Control Centre

When the ASTP mission was approaching, the Soviets disclosed that they were building an entirely new FCC for the joint spaceflight. Soyuz 12 was the first manned flight to be controlled from this new centre which was at first believed to be at Kalinin, 93 miles (150 km) northwest of Moscow. However, it was later established that the FCC was in Kaliningrad, 6 miles (10 km) northeast of Moscow and 6 miles (10 km) northwest of the Yuri Gagarin Cosmonaut Training Centre.

American reporters covering the ASTP mission were allowed to visit it in mid-May 1975, and were informed that construction had begun in 1970. One of the American ASTP Flight Directors, M.P. Frank, said that he was impressed by the quality of the Kaliningrad centre, but had the definite feeling that it did not have the performance capability of the American MCC.

The main operations room has 5 banks of consoles, 24 in total, with a large map of the World displayed on a screen at the front of the room. This displays the orbital path and the various tracking stations, and additional data is listed on panels at the side.

The original FCC is located at Yevpatoria, at the same site in the Crimea as the deep space tracking station. Having two control centres in operation allowed the Soviets to conduct the Soyuz 18/Salyut 4 mission at the same time as the ASTP flight. The ASTP Soyuz was controlled by Kaliningrad, and the Salyut mission by Yevpatoria.

The FCC is responsible for flight control, the determination of orbital parameters and the calculation of any manouevres necessary for changing the orbit, rendezvous with other spacecraft, and re-entry.

Specialist personnel include the medical officer, who thoroughly analyses the bio-telemetric data and assesses the physical condition of the crew; the **capsule communicator** who carries out voice exchanges with the crew; and the specialist responsible for the ground station network who monitors the operation of the stations, with which he maintains direct communication.

The Shift Flight Director summarises all the incoming data and makes decisions with regard to the further flight programme. If required, spacecraft mock-ups and simulators at the Yuri Gagarin Cosmonaut Training Centre can be used for identifying failures and working out corrective actions.

CapCom Assignments (Missions)

Mission	**CapComs**
Vostok 1	Popovich
Vostok 2	
Vostok 3	Gagarin
Vostok 4	Gagarin
Vostok 5	
Vostok 6	
Voskhod 1	
Voskhod 2	
Soyuz 1	Zudov
Soyuz 2	Unmanned
Soyuz 3	Zudov
Soyuz 4	
Soyuz 5	
Soyuz 6	Artyukhin, Zudov, *Glazkov, Kizim*
Soyuz 7	Artyukhin, Zudov, *Glazkov, Kizim*
Soyuz 8	Artyukhin, Zudov, *Glazkov, Kizim*
Soyuz 9	Artyukhin
Soyuz 10	
Soyuz 11	Artyukhin

Mission	CapComs
Soyuz 12	
Soyuz 13	
Soyuz 14	Glazkov, Gorbatko, Khrunov, Volynov
Soyuz 15	Glazkov
Soyuz 16	
Soyuz 17	Glazkov, Ryumin
April 5 anomaly	
Soyuz 18	Glazkov, Ryumin
Soyuz 19	Dzhanibekov, Illarionov, Romanenko, Shonin
Soyuz 20	Unmanned
Soyuz 21	Artyukhin, Popovich, Titov, G
Soyuz 22	Filipchenko
Soyuz 23	Artyukhin, Gorbatko
Soyuz 24	Artyukhin
Soyuz 25	
Soyuz 26	Glazkov, Rozhdestvensky, *Sevastyanov*
Soyuz 27	Glazkov
Soyuz 28	*Sevastyanov*
Soyuz 29	*Aksyonov*, Grechko, Rozhdestvensky
Soyuz 30	Gorbatko, Khrunov, *Sevastyanov*
Soyuz 31	Filipchenko, Rozhdestvensky
Soyuz 32	*Andreyev*, Grechko, Rozhdestvensky *Sevastyanov*
Soyuz 33	
Soyuz 34	Unmanned
Soyuz T	Unmanned
Soyuz 35	Rozhdestvensky
Soyuz 36	Rozhdestvensky

Soyuz T-2	
Soyuz 37	
Soyuz 38	Rozhdestvensky
Soyuz T-3	
Soyuz T-4	
Soyuz 39	Rozhdestvensky
Soyuz 40	Rozhdestvensky
Soyuz T-5	
Soyuz T-6	
Soyuz T-7	
Soyuz T-8	
Soyuz T-9	Rozhdestvensky
Launch abort	
Soyuz T-10	
Soyuz T-11	
Soyuz T-12	
Soyuz T-13	
Soyuz T-14	

CapCom Assignments (Cosmonauts)

Cosmonaut	**Mission**
Aksyonov	*Soyuz 29*
Alexandrov	
Andreyev	*Soyuz 32*
Anokhin	
Artyukhin	Soyuz 6, 7, 8, 9, 11, 21, 23, 24
Atkov	
Belyayev	

Cosmonaut	Mission
Beregovoi	
Berezovoi	
Bykovsky	
Demin	
Dobrovolsky	
Dzhanibekov	Soyuz 19
Feoktistov	
Filipchenko	Soyuz 22, 31
Gagarin	Vostok 3, 4
Glazkov	*Soyuz 6, 7, 8*, 14, 15, 17, 18, 26, 27
Gorbatko	Soyuz 14, 23, 30
Grechko	Soyuz 29, 32
Gubarev	
Illararionov	Soyuz 19
Ivanchenkov	
Kizim	*Soyuz 6, 7, 8*
Khrunov	Soyuz 14, 30 and Salyut 5
Klimuk	
Komarov	
Kovalyonok	
Kubasov	
Lazarev	
Lebedev	
Leonov	
Lyakhov	
Makarov	
Malyshev	

Nikolayev	
Patsayev	
Popov	
Popovich	Vostok 1, Soyuz 21
Romanenko	Soyuz 19
Rozhdestvensky	Soyuz 26, 29, 31, 32, 35, 36, 38, 39, 40, *T-9*
Rukavishnikov	
Ryumin	Soyuz 17, 18
Sarafanov	
Savinykh	
Savitskaya	
Serebrov	
Sevastyanov	*Soyuz 26, 28, 30, 32*
Shatalov	
Shonin	Soyuz 19 (Chief CapCom)
Solovyov	
Strekalov	
Tereshkova	
Titov, G	Soyuz 21
Titov, V	
Vasyutin	
Volk	
Volkov, A	
Volkov, V	
Volynov	Soyuz 14
Yegorov	
Yeliseyev	

Cosmonaut	Mission
Zholobov	
Zudov	Soyuz 1, 3, 6, 7, 8

1.22: Soviet Tracking Ships – 'The Cosmonauts.'

Introduction

The Soviet Union has a nationwide tracking system, and the great spread of their country (about 2½ times that of the United States) gives them a good systems start without leaving their own territory. However, there have been occasional press reports of Soviet efforts to put some tracking facilities into other countries such as in Africa or in Cuba. Their principal space tracking tracking facility is Yevpatoria in the Crimea, and others are located at Tbilisi, Dzhusaly, Kolpashevo, Ulan-Ude, Ussuriyisk, Petropavlovsk, and Medvezhyie Ozera.

Because there are certain constraints to their land-based tracking network, the Soviets have put considerable emphasis on developing a sea-based tracking support system, relying on ships to fill the gaps. The alternative would be a series of foreign land-based stations, as used by the United States, but this system has not been pursued probably because of Soviet reluctance to become too dependent upon foreign countries, or perhaps because not all nations approached were willing to be hosts.

Many of the tracking ships have been named in honour of dead cosmonauts, and one has been named in honour of Sergei Korolyov, the 'Chief Designer.'

Background

The need for the tracking ships is explained by the fact that spacecraft are only within direct radio-visibility from Soviet territory for about 9 hours out of 24.

One group of ships operates in the mid-Pacific, and serves to record military missile tests, in the area where the dummy warhead is to splash; or in sight of the orbital path of spacecraft overflying the Pacific, usually for their initial revolution.

Other ships have for some years operated in the tropical Atlantic and the Mediterranean along the paths of orbital flights. Such ships would put into various ports in these parts of the world for supplies and crew rest, and when they left port it was usually an indication that new space launches were pending.

The Pacific tracking ships appear to be military ships, whereas those in the Atlantic and Mediterranean belong to the Soviet Academy of Sciences. Eleven ships have been named as assigned to the Academy for space work, and often these are described as scattered around the world for space support purposes. They are as follows:-

Name	Displacement*
Cosmonaut Vladimir Komarov	17,500 tons
Academician Sergei Korolyov	21,465 tons
Cosmonaut Yuri Gagarin	45,000 tons
Cosmonaut Vladislav Volkov	9,000 tons
Cosmonaut Pavel Belyayev	9,000 tons
Cosmonaut Georgi Dobrovolsky	9,000 tons
Cosmonaut Viktor Patsayev	9,000 tons
Borovichi	9,000 tons
Kegostrov	9,000 tons
Morzhovets	9,000 tons
Nevel	9,000 tons

*Displacement = the weight fully loaded expressed in metric tons.

Cosmonaut Vladimir Komarov

Launched in 1966 as the dry cargo ship, the Genichesk, it was converted to its present role in 1967, and has 1000 berths. It has been used for lunar support work in the tropical waters of the Western Atlantic Ocean, and is frequently in Havana harbour.

Academician Sergei Korolyov

Launched in 1970. It was purpose-built, and can stay at sea for up to 120 days at a time.

Cosmonaut Yuri Gagarin

The largest purpose built tracking ship, launched in 1970. Its automatic systems allow it to pick up space communications even during Force 7 storms.

Cosmonaut Vladislav Volkov, Cosmonaut Pavel Belyayev, Cosmonaut Georgi Dobrovolsky, Cosmonaut Viktor Patsayev

The Cosmonaut Vladislav Volkov, first of four new tracking ships, entered service in 1977, followed by the Cosmonaut Pavel Belyayev, Cosmonaut Georgi Dobrovolsky, and Cosmonaut Viktor Patsayev, retiring some of the earlier ships. All are similar in appearance, and are designed to work efficiently in the roughest seas.

Operational Deployment

During most of the early Soyuz flights, one of the high capacity civilian ships anchored off Sable Island, Nova Scotia where four successive orbits passed within easy direct communication range. However in some later phases of the Salyut 6 mission, this station appeared to have been abandoned in favour of a new location in the Western Mediterranean close to Gibraltar. Another location connected with manned flights is in the Gulf of Guinea, West Africa, to monitor retrofire just before the re-entry and landing near Karaganda in the Soviet Union.

During the ASTP mission, a large tracking ship was located off the coast of Honduras to supplement the Sable Island position.

Soviet Molniya satellites relay information between tracking ships and the Soviet Union. The principle of launch constraints is to make flights only at times and places which will be in view of Soviet tracking stations at critical periods during the flight, and at other times to do without tracking, storing on tape in the spacecraft some data for later replay.

The radio-horizons of the various tracking stations within the Soviet Union and of the ships in their deployed positions are shown as ovals on the large world map display in the Kaliningrad control centre.

In 1982, the Soviets filed plans with the International Telecommunication Union for a satellite system which appears to be very similar to the US TDRSS (Tracking and Data Relay Satellite System).

Postscript

It is understood that a new tracking ship was under construction in 1984/85, to be named after one of the men who commanded the Strategic Rocket Forces – Marshall Nedelin.

1.23: Deceased Cosmonauts

Introduction

A number of cosmonauts have been killed during actual missions, as well as in ground accidents. Those announced are detailed below, but this list is almost certainly incomplete.

A highly placed Soviet space official indicated privately to his American counterpart in April 1974 that eight Soviet cosmonauts in training had been killed in aircraft or car accidents since 1960, before having any opportunity to make a spaceflight.

Deceased Cosmonauts

Name	Date of Death	Cause of Death
Belyayev	10 January 1970	Peritonitis, following operation to remove stomach ulcers.[1]
Dobrovolsky	30 June 1971	Soyuz 11 re-entry.
Gagarin	27 March 1968	Jet crash during training flight.
Komarov	24 April 1967	Soyuz 1 parachute failure.
Liem[2]	September 1981	Air crash.
Patsayev	30 June 1971	Soyuz 11 re-entry.
Volkov, V	30 June 1971	Soyuz 11 re-entry.

Notes

1 This is the official cause of death – but see biography.

2 Back-up Vietnamese cosmonaut.

1.24: Friendship Societies

Introduction

Many Soviet cosmonauts are used as political spokesmen for their country, to link their celebrity status with the political ideas and policies being advocated by the Soviet leadership. One method of doing this has been to appoint cosmonauts to positions on the boards of the various Friendship Societies established by the Soviet Union with many other countries. The following table lists the cosmonauts who are active in this field, and the position they hold within the named Society.

N.B. Only those cosmonauts who have actually flown in space, and thus achieved the necessary 'celebrity status,' are shown.

Friendship Societies

Cosmonaut	Society	Position
Aksyonov	—	
Alexandrov	—	
Artyukhin	Portugal	Vice-President

Cosmonaut	**Society**	**Position**
Atkov	—	
Belyayev	—	
Beregovoi	Hungary France	Vice-President President
Berezovoi	—	
Bykovsky	—	
Demin	—	
Dobrovolsky	—	
Dzhanibekov	Spain	Vice-President
Feoktistov	—	
Filipchenko	GDR	Vice-Chairman
Gagarin	Cuba	President [1]
Glazkov	Southern Yemen	Vice-President
Gorbatko	Mongolia	President
Grechko	—	
Gubarev	—	
Ivanchenkov	—	
Kizim	—	
Khrunov	—	
Klimuk	—	
Komarov	—	
Kovalyonok	—	
Kubasov	America	Vice-President
Lazarev	NDRI	Vice-President
Lebedev	—	
Leonov	Italy Egypt	Vice-President Deputy-Chairman
Lyakhov	Sri Lanka	Vice-President

Makarov	Ethiopia	Unknown [2]
Malyshev	Nepal	Vice-President
Nikolayev	—	
Patsayev	—	
Popov	Laos	Vice-President
Popovich	Austria	Vice-Chairman
Romanenko	Madagascar	Vice-President
Rozhdestvensky	Angola	Vice-President
Rukavishnikov	—	
Ryumin	—	
Sarafanov	—	
Savinykh	—	
Savitskaya	—	
Serebrov	—	
Sevastyanov	Finland	Vice-President
Shatalov	Cuba	President [1]
Shonin	—	
Solovyov	—	
Strekalov	—	
Tereshkova	Algeria	President
Titov, G	Vietnam	Chairman
Titov, V	—	
Vasyutin	—	
Volk	—	
Volkov, A	—	
Volkov, V	—	
Volynov	Poland	Vice-President
Yegorov	—	

Cosmonaut	Society	Position
Yeliseyev	—	
Zholobov	—	
Zudov	Ethiopia	Unknown [2]

Notes

1 Gagarin was President of the Soviet-Cuban Friendship Society until his death. Shatalov subsequently took over.

2 Makarov and Zudov are apparently both involved in the Soviet-Ethiopian Friendship Society, although in what capacity is unknown. It is the only Friendship Society which boasts two cosmonauts.

1.25: Time in Space

Introduction

The following table shows the cumulative time spent in space by each cosmonaut, expressed in days:hours:minutes.

Cosmonaut	Time in space
Valery Ryumin	361:21:33
Vladimir Lyakhov	324:11:22
Leonid Kizim	249:17:58
Viktor Savinykh	242:22:29
Vladimir Solovyov	236:22:50
Oleg Atkov	236:22:50
Valentin Lebedev	219:06:01
Vladimir Kovalyonok	216:10:11
Anatoly Berezovoi	211:09:05
Leonid Popov	200:14:45
Alexander Alexandrov	149:10:46
Alexander Ivanchenkov	147:12:39

Vladimir Dzhanibekov	145:15:59
Georgi Grechko	134:20:33
Yuri Romanenko	104:06:43
Vitaly Sevastyanov	80:16:19
Pyotr Klimuk	78:18:19
Vladimir Vasyutin	64:21:52
Alexander Volkov	64:21:52
Boris Volynov	52:07:17
Vitaly Zholobov	49:06:23
Alexei Gubarev	37:11:36
Viktor Gorbatko	30:12:48
Vladislav Volkov	28:17:02
Georgi Dobrovolsky	23:18:22
Viktor Patsayev	23:18:22
Gennady Strekalov	22:17:07 [1]
Andrian Nikolayev	21:15:21
Valery Bykovsky	20:17:47
Oleg Makarov	20:17:45
Svetlana Savitskaya	19:17:06
Valery Kubasov	18:18:00
Pavel Popovich	18:16:27
Yuri Glazkov	17:17:26
Yuri Artyukhin	15:17:30
Vladimir Aksyonov	11:20:11
Yuri Malyshev	11:20:00
Igor Volk	11:19:14
Anatoli Filipchenko	10:21:04
Alexander Serebrov	9:22:10

Cosmonaut	Time in space
Vladimir Shatalov	9:21:58
Nikolai Rukavishnikov	9:21:11
Alexei Yeliseyev	8:22:16
Vladimir Remek	7:22:16
Miroslaw Hermaszewski	7:22:03
Jean-Loup Chrétien	7:21:51
Rakesh Sharma	7:21:41
Sigmund Jähn	7:20:49
Bertalan Farkas	7:20:46
Arnaldo Tamayo-Méndez	7:20:43
Jugderdemidiin Gurragcha	7:20:43
Pham Tuan	7:20:42
Dumitru Prunariu	7:20:41
Patrick Baudry	7:01:38 [2]
Alexei Leonov	7:00:33
Georgi Shonin	4:22:43
Georgi Beregovoi	3:22:51
Valentina Tereshkova	2:22:50
Vladimir Komarov	2:03:05
Vladimir Titov	2:00:18 [1]
Gennady Sarafanov	2:00:12
Lev Demin	2:00:12
Vyacheslav Zudov	2:00:07
Valery Rozhdestvensky	2:00:07
Yevgeny Khrunov	1:23:39
Vasily Lazarev	1:23:38
Georgi Ivanov	1:23:01

Pavel Belyayev	1:02:02
Gherman Titov	1:01:18
Konstantin Feoktistov	1:00:17
Boris Yegorov	1:00:17
Yuri Gagarin	01:48

Notes

1 Excludes the duration of the Soyuz launch abort i.e. the intended Soyuz T-10 mission.

2 Patrick Baudry served as back-up on Soyuz T-6, but never flew onboard a Soviet spacecraft. He did, however, make a spaceflight onboard the American Shuttle mission 51-G.

1.26: Cosmonaut EVA's

Introduction

Alexei Leonov was the first man in the world to perform an EVA (Extra Vehicular Activity – or spacewalk) when he left the Voskhod 2 capsule on 18 March 1965 and spent 24 minutes in open space.

However, although they were first, the Soviets then lagged behind the Americans for many years. Following the Soyuz 4/5 EVA in January 1969, no further EVA was carried out until 1977 and the EVA by Romanenko and Grechko from Salyut 6 to examine the suspect docking port.

With the flights of the Salyut 6 and Salyut 7 space-stations, there came a new maturity, as more frequent and more daring EVA's were made, to carry out a wide variety of experiments and repairs to the Salyut stations.

With the benefit of this experience, EVA's are becoming more common, and it would seem that all the main crews on Salyut 6 and 7 have been EVA trained. Two spacesuits are kept onboard the Salyut at all times, and these are of a 'one-fit' design, adjustable to fit any cosmonaut. Improved spacesuits were used by the Soyuz T-13 crew based on recommendations made following earlier EVA work. These featured better leg and arm mobility, better lighting using head lamps, and improved access to suit controls.

To help prepare for EVA, cosmonauts train in a water tank at the Yuri Gagarin Cosmonaut Training Centre. It is 23m in diameter and 12m deep, and contains models of a Soyuz craft and a Salyut station. In this neutral buoyancy facility, cosmonauts wearing spacesuits can practice various EVA operations.

It is difficult to establish accurate times for EVA, but the figures given in the following tables are those supplied by official Soviet sources. They are thought to be equivalent to 'hatch opening to hatch closing.' For example, Alexei Leonov reportedly spent 12 minutes outside the spacecraft during an EVA lasting 24 minutes in total, while Yeliseyev and

Khrunov are thought to have spent 15 minutes transferring from Soyuz 5 to Soyuz 4, during an EVA totalling 1 hour. And on Soyuz 26, Grechko was reported by the Soviets to have spent 20 minutes outside the spacecraft, and 1 hour and 28 minutes in the 'near vacuum conditions of space.'

In addition, with the advent of more frequent EVA's, it is believed that the Soviets have recently taken to giving approximate EVA durations. Several EVA's have been reported as lasting 5 hours, and it seems unlikely that this is the exact duration. However, pending clarification, this figure has been used.

EVA Missions

Mission	EVA Cosmonaut	Purpose of EVA
Voskhod 2	Leonov	First man to perform EVA. No scientific research conducted. Propaganda stunt only.
Soyuz 4/5	Khrunov Yeliseyev	First transfer of crew from one spacecraft to another. Purpose, other than propaganda, remains unclear – all subsequent transfers were through internal hatches. *Zond related ?*
Soyuz 26	Grechko Romanenko	To check docking mechanism and carry out Medusa experiment.
Soyuz 29	Ivanchenkov Kovalyonok	To carry out Medusa experiment.
Soyuz 32	Ryumin Lyakhov	To cut free radio-telescope. (Unplanned.)
Soyuz T-5	Lebedev Berezovoi	To carry out Medusa experiment.
Soyuz T-9	Alexandrov Lyakhov	Installation of additional solar panels.[1]
Soyuz T-10	Solovyov Kizim	Extensive repairs to fuel system, and installation of additional solar panels.
Soyuz T-12	Savitskaya Dzhanibekov	First woman to perform EVA. A series of cutting, plating and welding experiments were carried out.[2]
Soyuz T-13	Savinykh Dzhanibekov	Installation of additional solar panels, Medusa experiment, and Halley's Comet dust collector.

Notes

1 Alexandrov and Lyakhov were not trained to instal the solar panels. Gennady Strekalov and Vladimir Titov were supposed to conduct a special repair mission, but their Soyuz T-8 spacecraft was unable to dock with the Salyut. A replacement mission, which would have been designated Soyuz T-10, also failed due to a launch abort.

2 Dzhanibekov had originally trained to repair the Salyut fuel line, but Kizim and Solovyov insisted on completing the repairs themselves.

Cosmonauts in order of first EVA

1: Alexei Leonov

2: Alexei Yeliseyev

3: Yevgeny Khrunov

4: Georgi Grechko

5: Yuri Romanenko

6: Alexander Ivanchenkov

7: Vladimir Kovalyonok

8: Valery Ryumin

9: Vladimir Lyakhov[1]

10: Valentin Lebedev

11: Anatoly Berezovoi

12: Alexander Alexandrov

13: Vladimir Solovyov[2]

14: Leonid Kizim[2]

15: Svetlana Savitskaya[2,3]

16: Vladimir Dzhanibekov[2]

17: Viktor Savinykh

Notes

1 Lyakhov was the first cosmonaut to make a second EVA.

2 It has been assumed that the flight-engineer was the first to leave the spacecraft, as was the normal Soviet practice on earlier missions. No confirmation is available however.

3 Savitskaya was the first woman to make an EVA.

Comparative time spent performing EVA's

	hrs.mins
Vladimir Solovyov	22.46[1]
Leonid Kizim	22.46[1]
Vladimir Dzhanibekov	08.35[2]
Vladimir Lyakhov	07.08[3]
Alexander Alexandrov	05.45[4]
Viktor Savinykh	05.00
Svetlana Savitskaya	03.35
Valentin Lebedev	02.33
Anatoly Berezovoi	02.33
Alexander Ivanchenkov	02.05
Vladimir Kovalyonok	02.05
Georgi Grechko	01.28
Yuri Romanenko	01.28
Valery Ryumin	01.23
Alexei Yeliseyev	01.00
Yevgeny Khrunov	01.00
Alexei Leonov	00.24
TOTAL:	91.34

Notes

1 Solovyov and Kizim performed 6 separate EVA's lasting:

04.15
04.56
02.45
02.45
03.05
05.00
22.46

2 Dzhanibekov performed 2 separate EVA's lasting:

03.35
05.00
08.35

3 Lyakhov performed 3 separate EVA's lasting:

01.23
02.50
02.55
07.08

4 Alexandrov performed 2 separate EVA's lasting:

02.50
02.55
05.45

Number of EVA's performed

Six

Leonid Kizim

Vladimir Solovyov

Three

Vladimir Lyakhov

Two

Alexander Alexandrov

Vladimir Dzhanibekov

One

Anatoly Berezovoi

Georgi Grechko

Alexander Ivanchenkov

Yevgeny Khrunov

Vladimir Kovalyonok

Valentin Lebedev

Alexei Leonov

Yuri Romanenko

Valery Ryumin

Viktor Savinykh

Svetlana Savitskaya

Alexei Yeliseyev

1.27: The Patronymic System

Introduction

Patronymic – derived from the name of its bearer's father.

The Soviet people do not have 'Christian' names as such. In their culture a special patronymic name is used in addition to the surname.

The first name is known as the imya, and the second as the otchestvo, or patronymic, which is made up of the father's imya plus the suffix -ich, -ovich, or -evich in the case of males, and -ovna or -evna in the case of females.

Example

Pyotr Ilyich Klimuk. His father's name was Ilya Fyodorovich Klimuk. The surname remained the same, a new imya – Pyotr – was chosen, and his father's imya became Klimuk's otchestvo – Ilyich.

In everyday usage, Klimuk would commonly be referred to as Pyotr Ilyich.

1.28: Ranks of the Soviet Army/Air Force/Navy

Introduction

On many occasions, a serving military cosmonaut has been promoted one rank upon returning from a successful mission e.g. from Captain to Major. The military ranks are as follows:-

Junior or 2nd Lieutenant

Senior or 1st Lieutenant

Captain

Major

Lieutenant-Colonel

Colonel

Major-General

Lieutenant-General

Colonel-General

General

Marshall

Background

Military conscription is compulsory in the Soviet Union – two years in the Army or three years in the Navy from the age of 18. (The Air Force is part of the Soviet Army.) However, military service can be deferred by going to University. Providing they graduate, conscripts then only have to do a basic 3 - 6 months as an officer before returning to their civilian career.

The time spent by conscripts in the Soviet forces involves military training, physical fitness, sports etc., and in addition, they also take part in young Communist and voluntary organisations' activities.

They are also given specialist training in trades that should prove useful in civilian life, including specialist trades such as radio-engineering, electronics, automation, remote control and power engineering.

When they have finished their military service, they are said to receive priority in admission to any university or institute of their choice.

1.29 : Soviet Honours and Awards

Introduction

Soviet cosmonauts traditionally receive the highest Soviet awards for bravery following their spaceflights – usually the title of Hero of the Soviet Union with accompanying Gold Star Medal, together with the Order of Lenin.

A second spaceflight merits the same awards, whereas a further mission normally results in the awarding of the Order of Lenin only.

In addition, the title of Pilot-Cosmonaut of the USSR is awarded upon completion of a first spaceflight.

N.B.
On two occasions, Soyuz 25 and Soyuz T-8, (both aborted after failed docking attempts) the normal awards were not made.

Background

The title of Hero of the Soviet Union is the Soviet Union's highest honour, and is awarded by the Presidium of the Supreme Soviet of the Soviet Union for a personal or collective feat of heroism.

The first seven men to receive the award were the airmen who rescued the crew and passengers of the Soviet steamer Chelyuskin squashed in the Arctic ice.

In 1933, as part of the Polar expedition led by Otto Schmidt, the ship was given the job of travelling along the Northern Sea route from Murmansk in the West to the Pacific port of Vladivostok. In February 1934, the steamer was jammed in the ice of the Chukchi Sea, so the crew and passengers disembarked and camped on the floes.

The seven fliers landed on crumbling ice floes to fly the shipwrecked people back to the mainland, a few at a time.

Originally, the Heroes wore no outward distinctions, but on 1 August 1939, a decree was issued instituting the Gold Star Medal to be awarded to Heroes of the Soviet Union. The five-pointed star, made of gold, weighed 21.5 grams.

Up to 1941, when the Nazis attacked the Soviet Union, 600 Heroes had been awarded. During the War years, more than 11,500 servicemen of many Soviet nationalities received it. Some foreigners who distinguished themselves in the War have also been awarded the title.

Since its inception in 1934, approximately 12,500 people have received the award, including nearly one hundred women. Individuals on whom the title is conferred also receive the Soviet Union's highest decoration, the Order of Lenin, introduced in April 1930.

For those awarded the title of Hero of the Soviet Union a second time, a bronze bust is erected in their honour at their place of birth.

1.30: Military and Civilian Salyut programmes

Introduction

It is generally agreed that in the period 1971 – 1977, the Soviets had two separate Salyut space-station programmes in operation: one optimised for civilian space experiments, the other for military applications. This section analyses the two different programmes, and examines the civilian/military aspects of the second-generation Salyuts which commenced operations in 1977.

Civilian Salyuts

The civilian Salyuts were Salyuts 1 and 4. It is also believed that Cosmos 557 would have been a civilian space-station had it succeeded. *(see section entitled 'Salyut 2/Cosmos 557 Crewing' for more details.)*

The civilian Salyuts were flown in higher orbits, which increased their value for astronomical observations. Telemetry was typical of Soviet non-military spacecraft, and although the crew commander was from the Soviet military, the flight-engineer was always a civilian.

Many photographs showing the interior of the civilian Salyuts have been released, as have detailed cut-away diagrams showing the interior and exterior appearance of the space-stations.

Military Salyuts

The military Salyuts were Salyuts 3 and 5. It is also believed that Salyut 2 would have been a military space-station had it succeeded. *(see section entitled 'Salyut 2/Cosmos 557 Crewing' for more details.)*

The military Salyuts were flown in lower orbits, presumably to get the most out of the capabilities of the onboard photo-reconnaissance equipment and other activities which replaced the astronomical observations carried out on the civilian Salyuts. These Salyuts used radio frequencies associated with other Soviet military space missions, and were conducted by all-military crews. They remained in orbit for shorter periods than their civilian counterparts and each ejected a capsule after the departure of the final crew. This was recovered on Earth, as is the practice with military reconnaissance satellites.

Only a very few photographs showing non-critical interior equipment onboard the military Salyuts have been released, and no detailed diagrams have been published. Therefore, the manner in which the data return capsules were attached to the Salyuts is not yet known.

Emergence of the Military programme

The separate military programme first became apparent with the launch of Salyut 3 in 1974. It was placed in a lower orbit than Salyut 1, and when the all-military crew entered the station, they switched to military telemetry.

A clear pattern emerged when the next space-station, Salyut 4, was plaçed in a similar orbit to Salyut 1. This time, the crew was again a mixture of military and civilian cosmonauts, and they remained on the civilian telemetry channels.

When Salyut 5 was launched, it paralleled the Salyut 3 mission profile and established the alternating civil-military-civil-military pattern of Salyut launches.

With hindsight, using the criteria of orbit, telemetry and launch sequence, it is possible to conclude that if they had been successful, Salyut 2 would have been the first military space-station and Cosmos 557 would have been the second civilian space-station.

The Soviets have never admitted to having two separate programmes and do not admit to using space for military purposes at all, much less to conducting military experiments on board their space-stations.

Salyut 6 and Salyut 7

Using the same criteria applied to the earlier space-stations, Salyuts 6 and 7 would be classified as civilian Salyuts, since the crews have been a mixture of military and civilian cosmonauts and civilian frequencies have been used. The stations remained operational for far longer than their predecessors, and their orbit of course varied, but most of the time was in the range used by Salyut 4.

A wide range of experiments were conducted, and these included astrophysical observations; Earth resources imaging; materials processing; and biological studies. A large number of experiments could be conducted on the long-duration flights, and some Western observers have speculated that military experiments were also carried out. Whether or not this was so, it is clear that the distinction which was drawn during the 1970's between military and civilian Salyuts can no longer be made, since if both types of activities are conducted, they are now carried out on the same space-station.

If one studies the wide range of Salyut 6 and 7 experiments that the Soviets have openly discussed, it is clear that the cosmonauts were kept extremely busy. If any military experiments were carried out, these could have only occupied a minimal amount of time.

It should be noted however, that there is a fine line between photographs of the Earth that are used for agricultural, geological and hydrological studies, and those that have military value. The most important characteristic is the spatial resolution of the images obtained. It is thought that the resolution of military reconnaissance satellites used for looking at specific objects may be as good as 2 in (5 cm), while those designed for broad area observations would probably require rather lower resolution, perhaps in the order of metres.

The Salyuts have been equipped with an MKF-6M camera, and its resolution is thought to be about 22 yards (20m), which is probably more suited to Earth resources observations than military reconnaissance. The other camera systems carried onboard are thought to have an even lower resolution than the MKF-6M.

However, it is possible that visual observations by the crew may be of greater military use. One potential military application is simply watching ships at sea to determine how easily they can be tracked. Another is observing the bioluminescence which is produced by plankton when it is disturbed, which may give clues to submarine locations.

In addition, some of the materials processing experiments may have had military applications. Cadmium-mercury-telluride, which was the subject of many of the materials processing experiments, can be used for military infra-red detectors, just as it can be used for civilian medical instruments, television components, or many other products.

Conclusions

So many experiments could have both civil and military applications that it is impossible to say that the Soviets have not conducted any military experiments onboard Salyuts 6 and 7.

It would appear however, that a separate military Salyut programme no longer exists. Certain military operations may have been turned over to automated spacecraft, and Soviet advancements in satellite photo-reconnaissance may have reduced the need for cosmonauts to gather data.

Part 2: Cosmonaut Biographies

Part 2: Cosmonaut Biographies

This part of the book contains comprehensive biographies and photographs of every Soviet and Intercosmos cosmonaut.

Because of the difficulties in obtaining up to date information, it should be noted that the ranks, wives occupations, number of children etc may be incorrect, and further studies e.g. Candidate of Science degrees, may have been omitted.

Biographies are also included for officially named Soviet cosmonauts who have either not yet flown, or have left the cosmonaut team before making a first spaceflight – i.e. Andreyev, Anokhin, Illarionov, Tortchillova and the 1960 cosmonauts. Other semi-official cosmonauts are included at the end of the main body of biographies.

No attempt has been made to include the many 'missing cosmonauts' whose names appeared in the early 1960's in various Western news reports. All were said to have been killed during a series of accidents on the launch pad or in orbit. Whilst the reports made exciting reading at the time, they have long since been discredited.

Anatoli, Dmitri, Grigori, Ivan, Mars, Valentin, Valentin 'Dyed' and Valentin Junior

The first group of Soviet cosmonauts was selected in 1960, and consisted of 20 men. Of those originally selected, only twelve subsequently made a spaceflight, the remaining eight having been removed from the cosmonaut programme during training. Only very brief details of these eight have been released, and they have been identified only by their imyas.

Most of the information about these men was released in Georgi Shonin's book entitled 'Samiye Pyerviye' (The Very First Ones), published in 1977.

Anatoli

Alexei Leonov has confirmed that Anatoli had to leave the cosmonaut team after some tests in the centrifuge at 12g. He believed the cause was extra-systoles (abnormal heart beats). In his book, Shonin also points to the centrifuge, but says that Anatoli developed haemorrhages in several parts of his body. The two reports are reasonably in agreement, and it would seem that Anatoli left the cosmonaut team in 1961.

Dmitri

Whilst many of those removed from the 1960 selection left comparitively early, Dmitri stayed in the team, and was assigned as back-up to Pavel Belyayev, commander of Voskhod 2, launched on 18 March 1965. Viktor Gorbatko was originally the back-up commander, but was replaced by Dmitri when he became ill.

Dmitri attended the Zhukovsky Military Air Academy and graduated in 1968, before being discharged from the cosmonaut team in 1969 on medical grounds.

In a group photograph of the Soyuz 6, 7 and 8 prime and back-up crews published in Shonin's book, an unknown cosmonaut bearing a strong resemblance to Dmitri is shown standing next to Yeliseyev. Although Shatalov also appears in the photograph, he is sitting at the front with Beregovoi, Kamanin and two Air Force Generals, whereas the prime and back-up cosmonauts are all standing in the back row.

From the photograph, it can be deduced that Dmitri was the original commander of Soyuz 8, but was replaced by Shatalov at a relatively late stage when he became disqualified on medical grounds. He left the cosmonaut team shortly afterwards.

Grigori

A Navy pilot, Grigori was originally included in the group of six men short-listed for the six Vostok missions. Together with Gagarin, Titov, Popovich, Nikolayev and Bykovsky, he went into intensive training.

In a 1961 photograph taken at Sochi, the six men posed with Korolyov and Dr. Karpov, the cosmonauts' first director of training, but in a later version of the same picture, Grigori had been carefully removed by retouching.

He served as the second back-up (i.e. support crew) to Yuri Gagarin on Vostok 1, launched on 12 April 1961, and as support crew for Andrian Nikolayev on Vostok 3, launched on 11 August 1962. It is believed that he should have flown onboard Vostok 4, but was removed from the programme, and was replaced by his back-up, Pavel Popovich.

In his book, Shonin's careful words suggest that Grigori had trouble adapting himself to the severe regime that governed both the work and leisure of the cosmonauts. He and other 'rakes' – Valentin Junior, Ivan and Mars – left in about 1962, obviously deemed to be unsuitable cosmonaut material.

Ivan

It has been speculated that he was the Ivan Anikeyev mentioned in Evgeny Riabchikov's book 'Russians in Space,' i.e. a room-mate of Titov's during the selection tests.

He was deemed unsuitable cosmonaut material – see Grigori.

Mars

A Tartar, he was deemed unsuitable cosmonaut material – see Grigori.

Valentin

He was said to be the most promising cosmonaut in the first group, and according to Shonin, was the only one who had no trouble at all with higher mathematics, part of the theoretical schooling of the team.

Unfortunately, one sunny day in 1960 or early 1961, he dived into a shallow lake near Star Town and damaged some of his lower dorsal vertebrae, spending a month in hospital. He apparently recovered nearly one hundred per cent, but was still medically discharged from the cosmonaut team.

Valentin 'Dyed'

He was one of the older cosmonauts, born in about 1925, while most of the others were born in the period 1933/35. Because of his appearance and his habits he was jokingly nicknamed 'Dyed' (Grandfather), to differentiate him from the two other Valentins in the group. He was said to be very fond of children, and when he had to leave the cosmonaut team at the beginning of 1962 (or a little later) due to medical reasons, the other cosmonauts were apparently very sorry to see him go.

In the book 'Test By Earth and Space,' by Pavel Popovich, published in 1982, it mentions that during the preparations for Voskhod 2, launched on 18 March 1965, Valentin 'Dyed' was a trainer to the crew in EVA technique. It therefore seems that instead of leaving Star Town upon his removal from the cosmonaut team, he stayed on as one of the many Air Force trainers.

Valentin Junior

Apparently known as Valentin Junior to differentiate him from the two other Valentins in the group, he was deemed unsuitable cosmonaut material – see Grigori.

Irina, Tanya and ?

When the Soviets decided to put a woman into space, a special cosmonaut group was selected in February 1962, consisting of four women. As well as Valentina Tereshkova, who flew the mission onboard Vostok 6, there were two others named simply as Irina and Tanya (otherwise Tatiana,) and a third unnamed woman, who was at one time thought to have been the back-up. All were said to be parachutists, but not all had pilot-training. Following the Vostok 6 mission, all three returned to their jobs, two as engineers in Star Town, while the other went to work in Moscow.

It was reported in 1981 however, that all four women had graduated from the Zhukovsky Military Air Academy in 1969. Of the three who did not fly, one was said to be a technical specialist, one a French teacher, and one worked at Star Town.

Over a period of years, some Western observers speculated that the unnamed woman thought to have been Tereshkova's back-up was called 'Vera,' or 'Venera.' However, it now seems that Tanya Tortchillova was the back-up, as in her memoirs, Tereshkova wrote:-

'Two female candidates were available, my friend Tanya and me. We both were fully trained in the operation of the Vostok systems. Why the Government Commission assigned me, I don't know. It was a last minute choice. A few female candidates were trained in the space centre and it was my sure and certain opinion that my best friend, Tanya Tortchillova was the prime candidate to pilot the Vostok 6. But finally, the choice was fixed upon me.'

Apparently, Tortchillova was married to Tereshkova's parachute instructor.

Therefore, it now seems that Tortchillova was the back-up to Tereshkova, with Irina or the unnamed woman as support.

The French cosmonauts actually met Irina and the other Vostok woman cosmonaut during their training at Star Town. Tortchillova was not present, which would appear to confirm the earlier report that one of the women went to work in Moscow following Vostok 6.

Irina

The Soviets have not yet revealed the back-up flight-engineer for Soyuz T-7, launched on 19 August 1982.

The prime flight-engineer was Svetlana Savitskaya, and some Western observers have since identified the Soyuz T-7 back-up crew, including Savitskaya's back-up, cosmonaut 'Irina'. No other details were forthcoming.

Irina was expected by these observers to fly as prime on the Soyuz T-12 mission, but in the event, Savitskaya made a second flight.

However, in a post-flight interview, Savitskaya stated that her back-up on Soyuz T-12 was 'a well trained specialist, a woman with a keen sense of purpose.' And, said Savitskaya 'I think she will have a chance to work in space.' Therefore, it appears that Irina may have served as back-up to Savitskaya for a second time.

Beregovoi told journalists in June 1982 that two women were undergoing training. Both were aged about 30, both were married, one was a pilot-engineer (presumably Savitskaya) and the other was a flight-engineer (presumably Irina.)

Vladimir Viktorovich Aksyonov

Vladimir Aksyonov was born on 1 February 1935, in the village of Giblitsy, in the wooded Oka Valley in the Kasimov District, Ryazan Region. His father was killed in World War Two, and his mother died when he was 14, so he was brought up by his grandparents. After finishing a seven-year school, he entered the Kasimov Industrial Technical School, but went on to the Mytishchi Engineering Technical School near Moscow, graduating in 1953. He then did a course at an Air Force School, and then attended the Chuguyev Air Force Officers School, graduating in 1957, having served for four years as a jet pilot.

Following graduation, he left the Air Force, and began work at Sergei Korolyov's design bureau, where he became Head of the Test Flying department. It is said that he was one of the workers involved with Sputnik 1. He became a senior design engineer, and was actively involved in the work of developing new spacecraft systems and in training cosmonauts for new spaceflights. As a test-pilot, he rehearsed practically everything that cosmonauts later did in orbit. He had to make himself at home in spacecraft, devise the methodology for the use of onboard instrumentation by cosmonauts, as well as preparing the necessary technical documentation. In addition, he also led sea-training exercises.

Also in 1957, he began a correspondence course with the All-Union Polytechnical Institute, where his attention was drawn to the design of life-support systems. He graduated from the Institute in 1963 with a Candidate of Technical Sciences degree. He joined the CPSU in 1959.

In 1966, Aksyonov worked with Valery Bykovsky to prepare the Soyuz 1/2 joint spaceflight. They worked together developing a system for transferring from one craft to another. The system was tested in airborne weightlessness, with Bykovsky representing the Cosmonaut Training Centre, and Aksyonov the design bureau. During these tests, Aksyonov logged a total of 9 hours of 'weightlessness'.

He was selected to join the cosmonaut team in 1973, and his first spaceflight came as flight-engineer onboard Soyuz 22, launched on 15 September 1976. During the mission, Aksyonov and Valery Bykovsky carried out extensive Earth resources photography, using a prototype of the East German MKF-6 multispectral camera, to be carried on future Salyut space-stations. They returned to Earth after a flight lasting 7 days 21 hours and 52 minutes.

Upon his return, Aksyonov was awarded the title of Hero of the Soviet Union and also received the Order of Lenin.

In 1978, he was involved in the ground control of Soyuz 29.

His second spaceflight came as flight-engineer onboard Soyuz T-2, launched on 5 June 1980. He and Yuri Malyshev docked with the orbiting Salyut 6 space-station and joined the crew of Popov and Ryumin already onboard. Their mission was the first manned test of the Soyuz T spacecraft, and the two men returned to Earth after a flight lasting 3 days 22 hours and 19 minutes.

Aksyonov was subsequently awarded the title of Hero of the Soviet Union and also received the Order of Lenin, both for the second time.

During the Soyuz T-7 and T-11 missions in 1982 and 1984 respectively, he acted as a radio commentator from FCC.

He is married, and his wife Marina Vasilyevna works in a design bureau. They have two sons, Valery and Sergei. Aksyonov likes poetry and music, is a Nature lover, and is the author of a biographical work about ex-cosmonaut Sergei Anokhin entitled 'The Hard Way', published in 1982. Although he is said to have left the Air Force, he is still referred to in some Soviet sources as a Colonel-Engineer.

Vladimir Aksyonov was the 36th Soviet cosmonaut, and he has spent a total of 11 days 20 hours and 11 minutes in space.

Captain-Engineer Alexander Alexandrov

NO PHOTOGRAPH AVAILABLE

Alexander Alexandrov was born on 1 December 1951, in the town of Omourtag, in Northern Bulgaria. His father Panayot Alexandrov is a forester.

Alexandrov studied at the Georgi Benkovski Higher Air Force School near Dolna Mitropoliya, not far from Preven. During his first or second year at the school, he crashed into a tree during a parachute jump and was badly injured. He was absent from studies for two to three months, but upon his return, he quickly caught up with his colleagues, and even managed to overtake many of them. He joined the Communist Party of Bulgaria in 1972, and graduated from the Higher Air Force School as a pilot-engineer in 1974. Upon graduation, he served in the anti-aircraft units of the Bulgarian Air Force.

Following the agreement to fly non-Soviet cosmonauts from member countries of the Intercosmos programme onboard Soyuz and Salyut spacecraft, Bulgaria began a cosmonaut selection programme. The two finalists were Alexandrov and Georgi Ivanov, and they were sent to begin their training at the Yuri Gagarin Cosmonaut Training Centre in March 1978. Ivanov eventually flew the mission as cosmonaut-researcher onboard Soyuz 33, launched on 10 April 1979, while Alexandrov served as back-up, and probably as CapCom in FCC.

Following the completion of the mission, Alexandrov was awarded the Order of the Peoples Republic of Bulgaria, First Class, for his 'persistence, conscientiousness and dedication during preparation and training for the flight.' He was also promoted from Lieutenant-Engineer to Captain-Engineer, awarded the title of Merited Pilot of the Peoples Republic of Bulgaria, and made an honorary citizen of Omourtag.

Alexandrov, known to his friends as Sasho (an abbreviation of Alexander) is not married. He has a keen sense of humour, and it said that he could have been an actor, but for his ambition to become a pilot. He is also reported to be an excellent trumpet player.

Alexander Pavlovich Alexandrov

Alexander Alexandrov was born on 20 February 1943, in Moscow. He joined the Soviet Army, and was attached to the elite Strategic Rocket Forces. However, he was disqualified for physical reasons in 1964, before completing his service. He then joined Sergei Korolyov's design bureau, where he took part in the development of spacecraft control systems. He combined work at the bureau with evening studies at the E.N. Bauman Moscow Higher Technical School, and graduated in 1969. He joined the CPSU in 1970.

He applied to join the cosmonaut team in 1965, but was told to wait, as he was too young. He therefore re-applied in 1967, but failed the centrifuge tests. He was finally selected to join the cosmonaut team in 1978, at which time he was working at FCC.

In 1979, he was one of the shift Flight Directors at FCC for the Lyakhov/Ryumin mission onboard Salyut 6, and he was on duty during the operation to untangle the radio-telescope antenna.

Alexandrov is known to have trained for a long-duration mission with Vladimir Dzhanibekov, and it is believed that his first assignment was as the flight-engineer on the support crew for Soyuz T-5, launched on 13 May 1982. (Dzhanibekov was subsequently moved to Soyuz T-6.) His next assignment was almost certainly as the back-up to Gennady Strekalov, the flight-engineer onboard Soyuz T-8, launched on 20 April 1983.

His first spaceflight came as flight-engineer onboard Soyuz T-9, launched on 27 June 1983. He and Vladimir Lyakhov docked with the orbiting Salyut 7 space-station, and settled down to a long programme of scientific experiments and observations.

A problem arose with the Salyut's solar panels, resulting in a reduced flow of electrical power. This seriously affected the environmental control system, resulting in cold, damp and uncomfortable conditions which threatened to make the station uninhabitable, and also posed a threat to the continued function of the station's electronics and other systems.

To overcome this serious problem, a repair crew of Vladimir Titov and Gennady Strekalov were assigned to train for a mission launched specifically to repair the faulty Salyut 7 solar array. However, following a docking failure, and then an aborted launch,

Lyakhov and Alexandrov, who had not been trained for such an operation, had to complete the repair themselves, by using detailed instructions read from the ground.

The repairs involved two separate EVA's, of 2 hours 50 minutes and then 2 hours 55 minutes, during which the two men had to fit two new solar panels sent up by a Progress spacecraft. The repairs were successful, and the crew resumed their normal programme. They returned to Earth after a flight lasting 149 days 10 hours and 46 minutes.

The Soyuz T-9 mission was originally planned to last approximately 100 days, but following Titov and Strekalov's launch abort, it was extended by 50 days.

Upon his return, Alexandrov was awarded the title of Hero of the Soviet Union and also received the Order of Lenin.

Known to his friends as Sasha, Alexandrov is married, and he and his wife Natalia met when they were both working at Korolyov's design bureau. She now works at FCC, and they have a son Sergei, born in 1976 or 1977.

Alexander Alexandrov was the 55th Soviet cosmonaut.

Boris Dmitrievich Andreyev

Boris Andreyev was born on 6 October 1940, in Moscow. After completing secondary school, he entered the E.N. Bauman Moscow Higher Technical School, where he specialized in automatic control systems. He was a keen athlete and also a member of the students scientific society.

Upon graduating in 1965, he began studying English, and went to work in the Korolyov design bureau. He began as an engineer, developing and testing elements of automatic control systems, and was then made chief of the scientific-technical information department. He joined the CPSU in 1969.

In 1970, he was selected to join the cosmonaut team, and in 1973, he was assigned to the Apollo-Soyuz Test Project (ASTP). His knowledge of English was probably a decisive factor in his selection. From May 1973 until July 1975, he served as back-up to Valery Kubasov, who flew as flight-engineer on Soyuz 19, the Soviet half of the joint mission, launched on 15 July 1975.

It is believed that following the completion of the ASTP mission, Andreyev may have trained for a Salyut flight in the late '70's, as did the other 1970 cosmonauts assigned to ASTP. Romanenko and Ivanchenkov remained paired when they were re-assigned to the Salyut programme, so it is likely that Dzhanibekov and Andreyev also stayed together. It is thought that they may have been the original Soyuz 25 support crew, but they were replaced when Andreyev became physically disqualified

All the 1970 cosmonauts assigned to ASTP have now made a spaceflight, with the exception of Valery Illarionov and Andreyev. In 1979, he was involved in the ground control of Soyuz 32, and he was still at Star Town in 1982. It is not known if he is still active.

He is married, and his wife Tamara Vasilyevna is also an engineer, having graduated from the E.N. Bauman Moscow Higher Technical School. They have a son Dmitri, and a daughter, Tatyana. Andreyev's hobbies include tinkering with his car, cycling, skiing and hiking.

Colonel Sergei Nikolayevich Anokhin

NO PHOTOGRAPH AVAILABLE

In his book, 'Touch Space', published in 1984, Valery Kubasov revealed that Sergei Anokhin was one of the engineers selected to join the cosmonaut team in 1966. However, after only a few months he was disqualified due to physical problems, probably due to his age, as he was 56 at the time of his selection.

Anokhin was a Hero of the Soviet Union, and was also a Merited test-pilot. He had been a member of Sergei Korolyov's design bureau, and it appears that he taught many of the engineer-cosmonauts of 1966 and later selections to fly.

Vladimir Aksyonov has written a biography of Anokhin entitled 'The Hard Way', published in 1982. In 1984, Anokhin was known to be serving as Sports Commissioner for Space Records.

Colonel-Engineer Yuri Petrovich Artyukhin

Yuri Artyukhin was born on 22 July 1930, in Pershutino, a small village near Moscow. After finishing secondary school in 1948, he entered the Serpukhov Air Force Technical School. He then served in various Air Force units for several years, before graduating from the Zhukovsky Military Air Academy in 1958. Following graduation, he started work in one of the Academy's research laboratories. He had joined the CPSU in 1957.

He was selected to join the cosmonaut team in 1963, and reported for duty on 11 January. In mid-1967, he was assigned to the Zond training group, and began preparing for a flight to the Moon. Upon the abandonment of the Soviet lunar programme, he was re-assigned to the Soyuz programme, and served as CapCom onboard the space tracking vessel 'Cosmonaut Vladimir Komarov' during the Soyuz 6/7/8 mission in 1969, and also for the Soyuz 9 flight in 1970. He was also CapCom for the Soyuz 11 mission in June 1971.

It is thought that Artyukhin may have trained for Salyut 2, but when that failed, he was re-assigned to Salyut 3. (see section entitled 'Salyut 2/Cosmos 557 Crewing' for more details.)

He had to wait 11 years from selection before making his first spaceflight, which came as flight-engineer onboard Soyuz 14, launched on 3 July 1974. He and Pavel Popovich had begun joint training for the mission in August 1973, and following launch, they docked with the orbiting Salyut 3 space-station and spent nearly two weeks onboard. They carried out a programme of Earth observations and bio-medical experiments before returning to Earth, after a flight lasting 15 days 17 hours and 30 minutes.

Upon his return, Artyukhin was awarded the title of Hero of the Soviet Union and also received the Order of Lenin. In 1976 and 1977 he served as CapCom for the Soyuz 21, 23 and 24 missions, and in February 1977, during the Soyuz 24 mission, he was also said to be a deputy flight controller. He was still known to be working at Star Town in 1981.

He is married, and his wife Nina Ivanovna works at the Nauka (Science) Publishing House. They have two sons, Vladimir, and Sergei, a former student of the Higher Aviation School in Riga. Artyukhin is Vice-President of the Soviet-Portugal Friendship Society.

Yuri Artyukhin was the 30th Soviet cosmonaut.

Oleg Yurievich Atkov

Oleg Atkov was born on 9 May 1949, in the village of Khvorostyanka in Khvorostyanka Rayon, Kuybyshev Oblast near the Volga. His father is a service man, and his mother a Russian-language teacher.

In 1973, Atkov graduated from the Moscow Sechenov No.1 Medical Institute, and he continued his education first as an intern and then as a research student at the All-Union Cardiological Research Centre of the Soviet Academy of Sciences.

A Candidate of Sciences (Medicine), he has specialised in ultrasonic investigation of the heart action, including ultrasonic methods of diagnosing heart disease.

He first became involved in space-medicine work in 1975, and was selected to join the cosmonaut team in 1977. He joined the CPSU in the same year. Although a cosmonaut, he remains a staff member of the Cardiological Research Centre.

He is credited with several inventions and a number of scientific works, and in 1978, he was awarded the Lenin Komsomol Prize as one of the inventors of a portable ultrasound cardiograph which was later used to examine Anatoli Berezovoi and Valentin Lebedev during their 211-day mission onboard Salyut 7 in 1982.

His first assignment came in September 1983, when he was assigned as research-cosmonaut on the Leonid Kizim and Vladimir Solovyov crew. The three men were launched onboard Soyuz T-10, on 8 February 1984, and docked their spacecraft with the orbiting Salyut 7 space-station. They carried out a long programme of research and experiments as well as playing host to the joint Soviet-Indian spaceflight Soyuz T-11, and the visiting Soyuz T-12.

Commenting on the decision to send a doctor into space, Georgi Beregovoi said 'It is useful when a spaceship has a specialist onboard who can directly observe the human organism's reaction to weightlessness and analyse in totality all the phenomena associated with it . . Oleg Atkov has experience in work with cosmonauts, he took part in medical check-ups of many crews and besides everything else, as a therapist he may be useful in spaceflight.'

Atkov was, of course, the second Soviet doctor in space – following Boris Yegorov's flight onboard Voskhod 1 some twenty years earlier.

According to Soviet reports, further progress with long stays in orbit required that more sophisticated medical research be carried out in space, rather than on Earth, and only a specialist could perform such experiments and evaluate and analyse the information thus received. 'In order to assess human reserves and potentialities,' said one report, 'it becomes necessary to place the cosmonauts under additional load. To do this in orbit without the presence of a doctor would be too much of a risk.'

Atkov's main studies onboard Salyut 7 were related to the cardio-vascular system and the process of salt and water exchange in the human body – data vital for prolonged spaceflights. It was expected that he would be able to considerably improve Soviet life science studies using himself and his colleagues as the research subjects.

A major task carried out by Kizim and Solovyov was a series of six EVA's during which they completed extensive and complicated repairs to the Salyut. Atkov remained onboard and supervised the EVA's.

The three men resumed their normal programme of research for a further two months, before returning to Earth after a flight lasting 236 days 22 hours and 50 minutes – thus setting a new duration record.

Upon his return, Atkov was awarded the title of Hero of the Soviet Union and also received the Order of Lenin. For his part in the joint Soviet-Indian mission, he was also awarded the Indian Order of Kirti Chakra.

He is married with children, and his wife is an eye surgeon.

Oleg Atkov was the 57th Soviet cosmonaut.

Lieutenant-Colonel Patrick Baudry

Patrick Baudry was born on 6 March 1946, in Douala, in the French colony of Cameroun, West Africa (now the United Republic of the Cameroons), where his father was based as a meteorological engineer. His family originally came from the Bordeaux region of France, but his parents, Mr and Mrs Roger Baudry, now live in Hyeres. After two years in advanced mathematics, (Mathématiques Supérieures, then Mathématiques Spéciales), he entered the French Air Force Academy in Salon de Provence in October 1967. He graduated in 1969 with a Master's degree in aeronautical engineering, and was awarded his fighter pilot/pilot-engineer wings in 1970.

From 1970 - 75, he served as a chase-pilot in the 1/11 'Roussillon' operational Jaguar squadron, flying F100 and Sepecat Jaguar strike fighters, and was assigned to various bases in France. He also completed numerous operational missions in several African countries.

In 1975, he applied for a position as a student test pilot and was sent to the Empire Test Pilot School (ETPS), Boscombe Down, England, under an exchange programme between English and French test-pilot schools. He graduated two years later, receiving the Patuxent Trophy awarded to the best student. He then served as a test-pilot at the Flight Test Centre at Brétigny-sur-Orge, where for two years, he was involved in numerous test programmes, including development of an auto-throttle system, for the Mirage III and V fighter series. He has now flown more than 4000 hours, including 3300 in jets, logged on 150 aircraft types, including 35 combat aircraft. The planes he has flown include the Mirage III, V, F1, 2000, F104, Jaguar, Harrier, Lightning, F100, F4, A8, T33, Hunter, Camberra, Mystere-4, and Vautour. He also holds an airline transport pilots licence.

In April 1979, the Soviet Union offered France the chance to fly a cosmonaut onboard a joint Soviet-French spaceflight, along the same lines as the agreement to fly non-Soviet cosmonauts from member countries of the Intercosmos programme. The offer was accepted, and France began a cosmonaut selection programme in September 1979. In March 1980, four candidates went to Moscow for three months of Russian language studies, and the two finalists, Baudry and Jean-Loup Chrétien, were named on 12 June 1980. They were then sent to begin their training at the Yuri Gagarin Cosmonaut Training Centre, arriving on 7 September 1980.

After a years training, Baudry was named in September 1981 as back-up for the flight. Chrétien eventually flew the mission as research-cosmonaut onboard Soyuz T-6, launched on 24 June 1982 while Baudry served as back-up, and also as CapCom in FCC.

Following the successful completion of the mission, Baudry was awarded the Soviet Order of People's Friendship, and the Soviet Order of Gagarin. He has also received the following awards and honours:

Chevalier de l'Ordre National du Mérite (the French National Order of Merit); Chevalier de la Légion d'Honneur (Order of the French Legion of Honour); and Titulaire de la Médaille de l'Aéronautique (the French Aeronautics Medal.)

After Soyuz T-6, he returned to France and CNES assigned him as prime crew member for the next flight opportunity.

Baudry, Anatoly Berezovoi and Jean-Loup Chrétien were nearly killed, on 4 June 1983, when the light aircraft Baudry was flying crashed shortly after take-off. The aircraft experienced a gradual loss of power minutes after take-off from an airport at Morlaix, France, but fortunately, Baudry managed to land in a field, and all three escaped serious injury.

In March 1984, during President Mitterand's visit to the United States, America and France agreed to fly a French astronaut onboard a March 1985 Shuttle mission. The flight opportunity had apparently been under discussion between NASA and CNES since 1983.

Baudry was selected as the prime payload specialist for Shuttle mission 51-E, and it was intended that he would continue the French echography cardiovascular monitoring system experiments begun on the earlier Salyut 7 flight, and also conduct a series of equilibrium and vertigo measurements. Chrétien was assigned as the back-up to Baudry, and the two men began training at the Johnson Space Centre in October 1984. They were also assigned to the French Hermes spacecraft review teams.

However, mission 51-E was cancelled only six days before the planned 7 March 1985 launch, due to technical problems with the TDRS-B satellite it was due to carry into orbit. The crew was re-assigned to fly mission 51-D, and as a consequence, Baudry was bumped by the McDonnell-Douglas payload specialist, Charles W. Walker, who had already been assigned the 51-D slot, and whose experiments had already been placed in the Discovery spacecraft, leaving insufficient room for the French equipment.

Baudry was therefore re-assigned to mission 51-G, launched on 17 June 1985. The crew consisted of commander Daniel Brandenstein, pilot John Creighton, mission specialists Shannon Lucid, Steven Nagel and John Fabian, and payload specialists Baudry and Sultan Salman Al-Saud.

Baudry completed the planned series of cardiovascular and sensorimotor system tests using himself, Lucid and Al-Saud as subjects. The experiments used ultrasound technology to measure changes in human blood flow and biochemical electronic sensors to measure muscular activity and changes in body posture, and were intended to expand upon the experiments performed on Salyut 7. Enhanced data was expected because of the capability on Shuttle flights to conduct tests shortly after launch, rather than having to wait until docking with the Salyut.

Some of the equipment was of the same design used in the Soviet spaceflight, but interfaces and packaging had to be developed for use in the Shuttle mid-deck area.

The Shuttle returned to Earth after a flight lasting 7 days 1 hour and 38 minutes.

Baudry is married, and his wife Claude is an architect and civil-engineer. She went with him to Star Town, and their daughter, Mélodie, born on 6 March 1976, speaks Russian fluently, and went to ballet school and kindergarten in Star Town.

He is 5ft 8in (1.72 m) tall with brown hair and brown eyes. He is fluent in both English and Russian, and his hobbies include skiing, motocross, shooting, wind-surfing, sky-diving, marathon running and squash, and he is interested in cars. He is also interested in the study and collection of wines, and is a dedicated connoisseur of Bordeaux wines. He collects labels from the best chateau wines, and whilst in Star Town, added to his collection. He is a member of several wine tasters confreries, including 'Chevaliers du Tastevin', 'Jurade de St. Emilion', and 'Confrerie du Bontemps'.

He is also a member of numerous pilots associations, including the Association of European Astronauts, and is a correspondent of the Air and Space Academy.

Although Baudry is a French Air Force officer, he is on detachment to CNES for his spaceflight activities. His first detachment began in 1980, and this was renewed in late 1984 for a further five years. He will remain available for future flights with the Shuttle (NASA), Salyut (Soviet Union) or Spacelab (ESA).

Baudry has the distinction of being the only **offical** non-Soviet Intercosmos back-up to fly as prime on a later manned spaceflight i.e. he is the only one to have actually made it into space. In addition, he and Chrétien are the only spacemen from outside the USA and Soviet Union to have trained on both countries' spacecraft.

Colonel Pavel Ivanovich Belyayev (1925 – 1970)

Pavel Belyayev was born on 26 June 1925, in the village of Chelishchevo, in the Volgoda Region. He became an expert skier, having to ski 3 miles (5km) to school every day, where his favourite subjects were physics and geography. Upon leaving school, he worked as a lathe-operator in a factory in Kamensk-Uralsk, before entering a Naval Fighter Pilots School in Yeish, and then an Air Force College in Moscow. He saw action as a fighter-pilot against the Japanese in World War Two, and following the end of the War, he stayed with his unit attached to the Soviet Navy in the Far East for a further 11 years. He joined the CPSU in 1949.

In 1956 he went to the Red Banner (now Gagarin) Air Force Academy, graduating in 1959. Whilst at the Academy, he was considered for the new cosmonaut team, and he is said to have passed the entrance tests easily, amazing examiners with his ability to endure long rides on the centrifuge. He was officially selected in 1960, and became the first commander of the cosmonaut team, until 1961, when he broke his ankle. It was a complex and painful break – a double fracture of the left leg just above the ankle – which he incurred during an awkward parachute landing. The fracture healed very slowly, and it looked at one stage as if Belyayev would be removed from the programme. However, with encouragement from Yuri Gagarin, he exercised his ankle with increasingly heavy lead weights, or dumbells, which he kept hidden under his bed. This weight-lifting therapy worked, and he felt fit enough to ask for permission to make a parachute jump. At first, the doctors refused, but they apparently relented when Gagarin offered to jump with him. The jump was a success, and Belyayev returned to the training programme, having been absent for a year.

He was put in charge of the cosmonauts selected in 1963, and made his only spaceflight as commander of Voskhod 2, launched on 18 March 1965. During the flight, Alexei Leonov became the first man to walk in space, and as commander, it was Belyayev's job to monitor Leonov's respiration and heartbeat. At the end of the mission, Belyayev had to make a manually-controlled re-entry following the failure of the automatic system. This meant that the two men had to make an additional orbit, and they returned to Earth after a flight lasting 1 day 2 hours and 2 minutes. The Voskhod capsule touched down above the snow-line in a forest near Perm, in the Urals, some 1250 miles (2000km) from the target zone. The capsule landed between two enormous fir trees in deep snow, and when the snow gave way, the spaceship became dangerously wedged between the two trees. The two men had to wait 2½ hours for the recovery helicopter, and then had to stay overnight at the touchdown spot.

Upon his return, Belyayev was awarded the title of Hero of the Soviet Union and also received the Order of Lenin.

On 23 April 1967, Vladimir Komarov was launched onboard Soyuz 1. He was brought back to Earth after just over a day in orbit, and was killed when the spacecraft's parachute tangled following re-entry. It is now clear, despite official Soviet denials at the time, that a second spacecraft, Soyuz 2, was to have been launched on 24 April. Following docking, an exchange of crews would have taken place.

The Soyuz 2 prime and back-up crews have since been identified by Western observers. The prime crew were Valery Bykovsky, Alexei Yeliseyev and Yevgeny Khrunov, and the back-up crew were Pavel Belyayev/Andrian Nikolayev, Valery Kubasov and Viktor Gorbatko. (There is some uncertainty as to whether Belyayev or Nikolayev was the back-up Soyuz 2 commander.) Following the initial problems with Soyuz 1, the Soyuz 2 launch was cancelled.

An unmanned Soyuz 2 subsequently flew in a rendezvous operation with Soyuz 3, piloted by Georgi Beregovoi, and in January 1969, nearly two years late, the mission intended for Soyuz 1/2 was carried out by the Soyuz 4/5 spacecraft. The prime and back-up crews of Soyuz 2 were given the same assignments on Soyuz 5 – only the commanders had changed. The position of back-up commander was filled by Georgi Shonin.

Many of the rumours and reports emanating from Moscow following the Soyuz 1 launch named Belyayev and Bykovsky as members of the crew of the Soyuz 2 spacecraft. It has therefore been speculated that Belyayev, and not Nikolayev, was the back-up commander for Soyuz 2, although only circumstantial evidence exists.

Belyayev was put in charge of the induction training of the 1967 group of cosmonauts in May 1967 i.e. one month after the Soyuz 2 mission was cancelled. In addition, he began training for a Zond circumlunar flight in late 1967. The flight was probably scheduled for late 1968, and therefore his position in the Soyuz 5 back-up crew would have had to have been given to someone else.

Belyayev joined the Zond training group in late 1967 as training manager, replacing Vitaly Sevastyanov, and began preparing for a flight to the Moon. In a meeting at the Paris Air Show in early 1968, he had a conversation with American astronauts Michael Collins and David Scott, in which he told them that he expected to make a circumlunar mission in the 'not too distant future.' It was an off-the-record remark, and not a propaganda statement, which gives the remark a certain credibility. According to one Western report, the Zond mission was scheduled for early December 1968, but was cancelled at the last moment.

Sergei Korolyov once said of him 'He is calm, unruffled, I would even say a trifle slow, but very thorough. He is not given to making speeches, but everything he does is done thoroughly.' Others described him as 'Reserved, taciturn, self-controlled, and very well organised.' His hobbies included painting and writing poetry.

In December 1969, he became seriously ill and was hospitalised. According to one Soviet source, the doctors recommended an operation, as his condition was steadily deteriorating, but they were not optimistic of his chances of survival, as 'the disease was progressing.' This report, which seems to indicate cancer, conflicts with all other Soviet reports which stated that the purpose of the operation was to remove stomach ulcers.

Unfortunately, following the operation, he developed Peritonitis, and as a result, died on 10 January 1970. He left a widow, Tatyana (Tanya) Filippovna, and two children, Irina, born in 1949, and Ludmilla, born in 1955.

In his honour, the Soviets named a crater on the far side of the Moon after him, and also named one of the space tracking vessels the 'Cosmonaut Pavel Belyayev.' In addition, in October 1978, a Soviet-discovered asteroid 6 miles (10km) in diameter orbiting the Sun between Mars and Jupiter was officially named 'Belyayev.'

Alexei Leonov and Belyayev were great friends, and Leonov was to say later in tribute 'Pavel Belyayev was a man I sought to emulate. I learnt a great deal from him.'

Pavel Belyayev was the 10th Soviet cosmonaut.

Lieutenant-General Georgi Timofeyevich Beregovoi

Georgi Beregovoi was born on 15 April 1921, in Fyodorovka, a village in the Poltava Region of the Ukraine. Upon finishing school, he worked at a steel plant, and in his spare time, enjoyed making model aircraft as a hobby. After completing training at the Yenakiyevo flying club, he went to the Lugansk Air Force Pilots School, and joined the Soviet Air Force in 1938.

During World War 2, he flew 185 combat missions, and was decorated with many awards and honours, including in October 1944, the title of Hero of the Soviet Union, with the Order of Lenin. He joined the CPSU in 1943. When the War ended, he became one of Russia's leading test-pilots, flew in more than 50 types of aircraft, and was the Chief Test Pilot on all-weather interceptors. He went to the Red Banner (now Gagarin) Air Force Academy, and graduated in 1956.

Beregovoi was selected to join the cosmonaut team in 1964 at the age of 43, having taken part in more than 4000 test-flights. 'Even though he was over 40, I advised him to transfer to us. I promised him that in the next few years we would have work of interest to him . . . Testing new machines seems to be his element. Nature has endowed him with a good mind, nerves of steel, willpower and talent. A talent for test-flying.' said Sergei Korolyov. 'Our detachment has no technically better trained pilot that Georgi' Gagarin once said of him.

He reported for duty in February 1964, having been specially selected on the personal recommendation of Korolyov and General Kamanin. It was understood for many years that he was a solo 1964 selection, but in 1985, he revealed that he was in fact part of the second (1963) selection, and was simply a late arrival.

According to Beregovoi himself, he was found to be too fat and not strong enough for the rigours of spaceflight when he first arrived at the Yuri Gagarin Cosmonaut Training Centre. However, as he had all the other qualities needed, the centre's head of physical training, Yuri Surinov, put him on an intensive six-month course.

At the beginning of the course, Beregovoi took over 4 minutes to swim 100 metres, and over 8 minutes to run 1500 metres. At the end of the six months, he had reduced these times to 2 minutes 25 seconds and 5 minutes 32 seconds, respectively.

And in the gymnasium, the rather podgy and underdeveloped trainee cosmonaut apparently suprised his trainer by displaying a very high degree of physical co-ordination, which allowed him to master the apparatus in a phenomenally short time. By the end of the course, he had lost nearly a stone in weight, and was in the tip-top physical condition needed for a spaceflight.

It has been revealed that a *third* mission in the Voskhod series was planned for the early summer of 1966, involving a two-man team of cosmonauts in a long-duration mission. A training group was formed, probably around April 1965, but the mission was cancelled in early 1966.

The four cosmonauts since named as being part of this group were Beregovoi, Shatalov, Volynov and Shonin. It is thought that the prime crew would have been Volynov and Shonin, with Beregovoi and Shatalov as the back-ups. Following cancellation of the remaining Voskhod missions, the four men were re-assigned to the new Soyuz programme.

Beregovoi made his only spaceflight as commander of Soyuz 3, launched on 26 October 1968. It was the first manned test-flight of the Soyuz spacecraft since the Soyuz 1 accident when Vladimir Komarov was killed. Soyuz 3 carried out a rendezvous with the unmanned Soyuz 2 spacecraft, but did not dock with it, as had been expected by many Western observers. Beregovoi returned to Earth after a flight lasting 3 days, 22 hours and 51 minutes.

At 47, he had been the oldest man ever to have made a first spaceflight. (This record stood until August 1974, and Lev Demin's Soyuz 15 flight at the age of 48.) However, after the flight, he revealed that his height of 5ft 11in (1.80 m) had proved more of a problem than his age when it came to being selected for Soyuz 3.

He was subsequently awarded the title of Hero of the Soviet Union and also received the Order of Lenin, both for the second time.

Beregovoi served as a deputy to the Council of the Union of the Supreme Soviet from 1970-1974, and was re-elected in 1974. He is also Vice-President of the Soviet-Hungarian Friendship Society, and President of the Soviet-French Friendship Society.

He is currently Head of the Yuri Gagarin Cosmonaut Training Centre, and is believed to have taken up this position shortly after the Soyuz 3 mission. He was promoted to Lieutenant-General in 1977, and has a degree as a Candidate of Psychological Sciences.

Known to his friends as 'Zhora', he is married, and he and his wife Lydia Matreeva have two children, Lyudmilla and Viktor. Beregovoi has been described as 'A bear-like Ukrainian, with a vice-like handshake and ready smile.'

He once revealed that '26 October is a very special date for me. On that date in 1944, I received the Hero of the Soviet Union award for combat service during World War 2. I flew Soyuz 3 in 1968, and later on the same date received further awards. It is my lucky number!'

Georgi Beregovoi was the 12th Soviet cosmonaut.

Colonel Anatoly Nikolayevich Berezovoi

Anatoly Berezovoi was born on 11 April 1942, into a peasant family in the village of Znem, in the Oktyabrskiy District of the Adigei Autonomous Region, in the North Caucasus.

After completing secondary school, he worked as a lathe-operator in the Neftemash factory in Novocherkassk, a town in the Rostov Region in the Russian Federation. He has said that he first decided to become a cosmonaut following Yuri Gagarin's spaceflight, and realised that the first step was to enrol at an Air Force School, which he did in 1961. He graduated from the Kachinsk Military High School for Pilots in 1965, and then served with various Army Air units, achieving the qualification of Pilot, 1st Class. He joined the CPSU in 1966.

Berezovoi was selected to join the cosmonaut team in May 1970. *It is believed that his first assignment may have been as commander of the support crew for Soyuz 23, launched on 14 October 1976. (see section entitled 'Salyut 5 Crewing' for more details.) This speculation is based on Berezovoi's account of how he found himself at the bottom in a group of crew selections in April 1975, with no chance of flying. Following the failure of Soyuz 23, Berezovoi was moved up to the back-up position on the replacement mission.* His next assignment therefore was as the back-up to Viktor Gorbatko, the commander of Soyuz 24, launched on 7 February 1977. In the same year, he graduated from the Gagarin Air Force Academy, whilst continuing his work at the Yuri Gagarin Cosmonaut Training Centre.

He was paired with Valentin Lebedev in late 1980, and made his first spaceflight as commander of Soyuz T-5, launched on 13 May 1982. He and Lebedev docked their spacecraft with the orbiting Salyut 7 space-station and carried out a long programme of research and experiments as well as playing host to the joint Soviet-French spaceflight Soyuz T-6, and the visiting Soyuz T-7.

During their stay onboard the Salyut, Berezovoi and Lebedev made an EVA lasting 2 hours and 33 minutes. Berezovoi stayed in the hatchway while Lebedev removed samples from the Medusa experiment and replaced them with fresh materials. The two men also tested new tools and assembly procedures.

They returned to Earth after a flight lasting 211 days 9 hours and 5 minutes – thus setting a new duration record. Upon his return, Berezovoi was awarded the title of Hero of the Soviet Union and also received the Order of Lenin. In addition, he was made a member of the French Legion of Honour.

His next assignment came in October 1983, when he was named as back-up to Yuri Malyshev, the commander of the joint Soviet-Indian mission, Soyuz T-11, launched on 2 April 1984. For his part in the preparations for the mission, he was awarded India's Kirti Chakra Order.

He was nearly killed on 4 June 1983, when he and Jean-Loup Chrétien were passengers onboard a light aircraft piloted by Patrick Baudry, which crashed shortly after take-off. The aircraft experienced a gradual loss of power minutes after take-off from an airport at Morlaix, France, but Baudry managed to land in a field, and all three escaped serious injury.

Berezovoi, known to his friends as Tolya, is married, and he and his wife Lydia, a teacher at the Moscow College of Civil Engineering, have a son born in 1969 or 1970, and a daughter, Tatiana, born on 14 July 1974.

Anatoly Berezovoi was the 51st Soviet cosmonaut.

Colonel Valery Fyodorovich Bykovsky

Valery Bykovsky was born on 2 August 1934, in Pavlovsky Posad, a small suburb of Moscow, the son of a transport worker. As a child, he wanted to become a sailor, but at the age of 17, became interested in flying. He soon graduated from a flying club and then from the Kachinskoye Military Aviation School in 1955. He joined the Soviet Air Force and became a jet fighter pilot and parachute instructor. His father said of his career 'He has always been courageous, and exciting and dangerous professions attracted him.' A report by his commanding officers described him as a 'bold, intelligent pilot,' who was 'calm in flight, and makes fast decisions in complex situations.'

In 1960, he became one of the first to join the cosmonaut team, and he volunteered to try out the original Soviet space-training devices, including the isolation chamber. This was at a time when no-one really knew what effect prolonged tests in an isolation chamber would have on a man. In recognition of his work, Bykovsky was awarded the Order of the Red Star. He was very popular with his fellow cosmonauts, and they described his as 'a jolly good fellow, the general favourite.' His first assignment was as the back-up to Andrian Nikolayev, the pilot of Vostok 3 launched on 11 August 1962.

Bykovsky did not belong to the CPSU at the time of his selection, and only applied to join shortly before his first spaceflight, which came as pilot of Vostok 5, launched on 14 June 1963. He learnt during the flight that his application had been accepted. He spent two days in orbit before being joined by Valentina Tereshkova in Vostok 6, and during this, the world's second joint flight, the two cosmonauts regularly relayed TV pictures to Earth, and conversed with each other. Bykovsky returned to Earth after a flight lasting 4 days 23 hours and 6 minutes, setting a new duration record which stood for two years before being broken by the U.S. Gemini 5 crew.

Upon his return, Bykovsky was awarded the title of Hero of the Soviet Union and also received the Order of Lenin. In 1966, Bykovsky worked with Vladimir Aksyonov to prepare the Soyuz 1/2 joint spaceflight. They worked together developing a system for transferring from one craft to another. The system was tested in airborne weightlessness, with Bykovsky representing the Cosmonaut Training Centre, and Aksyonov the design bureau.

On 23 April 1967, Vladimir Komarov was launched onboard Soyuz 1. He was brought back to Earth after just over a day in orbit, and was killed when the spacecraft's parachute tangled following re-entry. It is now clear, despite official Soviet denials at the time, that a second spacecraft, Soyuz 2, was to have been launched on 24 April. Following docking, an exchange of crews would have taken place.

The Soyuz 2 prime and back-up crews have since been identified by Western observers. The prime crew were Bykovsky, Alexei Yeliseyev and Yevgeny Khrunov, and the back-up crew were Pavel Belyayev/Andrian Nikolayev, Valery Kubasov and Viktor Gorbatko. However, following the initial problems with Soyuz 1, the Soyuz 2 launch was cancelled.

An unmanned Soyuz 2 subsequently flew in a rendezvous operation with Soyuz 3, piloted by Georgi Beregovoi, and in January 1969, nearly two years late, the mission intended for Soyuz 1/2 was carried out by the Soyuz 4/5 spacecraft. The prime and back-up crews of Soyuz 2 were given the same assignments on Soyuz 5 – only the commanders had changed. Bykovsky was not available to command Soyuz 5, as in mid-1967, he was assigned to the Zond training group, and began preparing for a flight to the Moon.

In 1968, he graduated from the Zhukovsky Military Air Academy, and upon the abandonment of the Soviet lunar programme, he was re-assigned as Training Manager at the Yuri Gagarin Cosmonaut Training Centre, taking part in the preparation of all Soyuz flights. In 1973, he completed a thesis on stellar navigation, which earned him a degree as a Candidate of Technical Sciences, and he has since written several papers on this subject.

By 1973, he was publicly denying the rumours that he was 'no longer active.' *Some Western observers have speculated that he was the back up to Pyotr Klimuk, the commander of Soyuz 13, launched on 18 December 1973. However, as yet, there appears to be no evidence to support this.* In April 1974, he was given special responsibility for training the Soviet ASTP crews.

He was actually placed back on the active list in 1975, when he was replaced as Training Manager by Vasily Lazarev. Bykovsky began training for a new spaceflight in January 1976, and flew as commander of Soyuz 22, launched on 15 September, 1976. During the mission, Bykovsky and Vladimir Aksyonov carried out extensive Earth resources photography, using a prototype of the East German MKF-6 multispectral camera, to be carried on future Salyut space-stations. They returned to Earth after a flight lasting 7 days 21 hours and 52 minutes.

Bykovsky was subsequently awarded the title of Hero of the Soviet Union and also received the Order of Lenin, both for the second time. In May 1977, he was paired with Sigmund Jähn from the German Democratic Republic, and made a third spaceflight as commander of the joint Soviet-GDR Soyuz 31 mission, launched on 26 August 1978. He and Jähn docked with the orbiting Salyut 6 space-station, and joined the crew of Kovalyonok and Ivanchenkov already onboard. The two men carried out a joint programme of Soviet-GDR experiments and observations before returning to Earth after a flight lasting 7 days 20 hours and 49 minutes.

Bykovsky was subsequently awarded the Order of Lenin for the third time. In addition, he was awarded the Karl Marx Order of the GDR; the title of Hero of the GDR; the freedom of Berlin; the F.Gekkert Gold Medal of the GDR; the GDR Academy of Science's Leibnitz medal; and the Arthur Becker Gold Medal of the FDJ (the GDR's socialist youth organisation).

His next assignment came in September 1979, when he was paired with Bui Thanh Liem from Vietnam, and served as the back-up to Viktor Gorbatko, the commander of the joint Soviet-Vietnamese Soyuz 37 flight, launched on 23 July 1980. Under the crewing policy used by the Soviets for the Intercosmos programme, he should have flown as commander of the joint Soviet-Romanian Soyuz 40 flight, launched on 14 May 1981. In the event however, Bykovsky apparently became disqualified, and the back-up crew of Leonid Popov and Dumitru Prunariu flew the mission.

Bykovsky has a fiery temperament like Titov, who once described him as 'like a boy, very hot, and curbs his desires only with the greatest difficulty.' He was so moved when his wife gave birth to their first son that he wept. A colleague once said of him 'Pensiveness, not dreaminess is typical of Bykovsksy. He will suddenly abandon a conversation and become engrossed in his own thoughts. Or he suddenly becomes abrupt and just as suddenly returns to his customary politeness. He can read for hours without betraying his presence. Sometimes he seems shy, unsociable and self-centred. Even we, his friends, occasionally thought this of him.'

His wife Valentina Mikhailovna works in a museum in Star Town, and they have two children, Valery, born in 1963, and Sergei, born in 1965. Bykovsky is interested in politics, and also enjoys sports. His favourite game is gorodki, a form of bowling, but he also likes soccer, basketball and hockey. He was once asked what he liked doing most, to which he replied 'Being busy!'

Valery Bykovsky was the 5th Soviet cosmonaut, and he has spent a total of 20 days 17 hours and 47 minutes in space.

Colonel Jean-Loup Chrétien

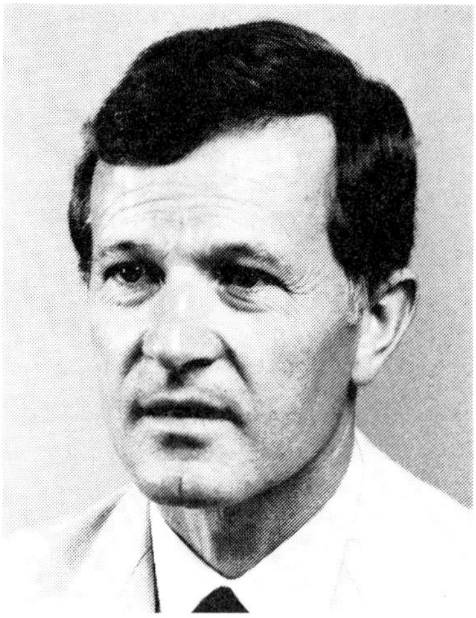

Jean-Loup Chrétien was born on 20 August 1938, in the town of La Rochelle, in France, and now lives in Morlaix, in Northern Brittany. His childhood dream was to become a pilot, and he began flying at a local aeroclub. After two years in advanced Mathematics, he entered the Air Force Academy at Salon de Provence in 1959, and graduated in 1961, receiving a Master's degree in aeronautical engineering.

He later recalled that following the flight of Vostok 1 in 1961, 'I asked what were the makings of a cosmonaut, but all would laugh at the question, and my father's friends said I'd be an old man when the first French cosmonaut flew into outer space.....Ever since Yuri Gagarin made his spaceflight, I wanted to become a cosmonaut, no matter what.'

He got his fighter pilot/pilot-engineer wings in 1962, after one years training on Mystere-4's. He was promoted to Lieutenant, and then joined the 5th Fighter Squadron in Orange, in the Southeast of France, where he served for seven years as a fighter-pilot in an operational squadron flying Mirage III interceptors. In 1970, he was assigned to the French test pilots school, EPNER (Ecole du Personnel Navigant d'Essais et de Réception), then served as a test-pilot at the Istres Flight Test Centre for seven years.

It is said that his flying skills, analytical powers, good discipline and commanding abilities were quickly noticed, and he was made responsible for supervising the flight-test programme for the Mirage F-1 fighter. In 1977/78, he was appointed Deputy Commander of the South Air Defence Division in Aix en Provence, and he served in this position until his selection as a cosmonaut. He has accumulated over 6000 hours of flying time.

In April 1979, the Soviet Union offered France the chance to fly a cosmonaut onboard a joint Soviet-French spaceflight, along the same lines as the agreement to fly non-Soviet cosmonauts from member countries of the Intercosmos programme. The offer was accepted, and France began a cosmonaut selection programme in September 1979. In March 1980, four candidates went to Moscow for three months of Russian language studies, and the two finalists, Chrétien and Patrick Baudry were named on 12 June 1980. Chrétien had apparently been one of the original candidates considered for the first Spacelab mission. The two men were sent to begin their training at the Yuri Gagarin Cosmonaut Training Centre, arriving on 7 September 1980.

After a years training, Chrétien was named in September 1981 as a member of the prime crew for the mission, and subsequently served as research-cosmonaut onboard Soyuz T-6, launched on 24 June 1982. Chrétien, Vladimir Dzhanibekov and Alexander Ivanchenkov linked up with Salyut 7, and joined the crew of Berezovoi and Lebedev already onboard. They spent nearly seven days carrying out a programme of joint Soviet-French experiments, including a series of French echography cardiovascular monitoring system experiments, before returning to Earth after a flight lasting 7 days 21 hours and 51 minutes.

During the mission, Chrétien displayed his sense of humour by donning a Quasimodo mask and suprising Valentin Lebedev with it, much to the amusement of the other cosmonauts !

Upon his return, he was awarded the title of Hero of the Soviet Union and also received the Order of Lenin. He has also received the following awards and honours:

Chevalier de l'Ordre National du Mérite (the French National Order of Merit); Chevalier de la Légion d'Honneur (Order of the French Legion of Honour); and Titulaire de la Médaille de l'Aéronautique (the French Aeronautics Medal.)

Chrétien, Patrick Baudry and Anatoly Berezovoi were nearly killed on 4 June 1983, when the light aircraft Baudry was flying crashed shortly after take-off. The aircraft experienced a gradual loss of power minutes after take-off from an airport at Morlaix, France, but fortunately, Baudry managed to land in a field, and all three escaped serious injury.

In March 1984, during President Mitterand's visit to the United States, America and France agreed to fly a French astronaut onboard a March 1985 Shuttle mission. The flight opportunity had apparently been under discussion between NASA and CNES since 1983.

Baudry was selected as the prime payload specialist for Shuttle mission 51-E, and Chrétien was assigned as back-up. The two men began training at the Johnson Space Centre in October 1984, and were also assigned to the French Hermes spacecraft review teams.

Unfortunately, mission 51-E was cancelled only six days before the planned 7 March 1985 launch, due to technical problems with the TDRS-B satellite it was due to carry into orbit. The crew was re-assigned to fly mission 51-D, and as a consequence, Baudry was bumped by the McDonnell-Douglas payload specialist, Charles W. Walker, who had already been assigned the 51-D slot, and whose experiments had already been placed in the Discovery spacecraft, leaving insufficient room for the French equipment.

Baudry and Chrétien were therefore re-assigned to mission 51-G, launched on 17 June 1985.

Chrétien is married, and has four sons – Jean-Baptiste, born in 1962, Olivier, born in 1965, Emmanuel, born in 1966 and Francois, born in 1974. His family stayed in France and did not go with him to Star Town during his training.

He is 6 ft (1.82 m) tall, and has blonde hair and blue eyes. He is fluent in English and Russian, and his hobbies include skiing in Winter and sailing in Summer. He also enjoys wind-surfing, car-rallying and woodworking. In addition, he plays the electric organ, and took it with him to Star Town. He is fond of music, particularly classical, and his favourite singers are Guy Béart, Georges Brassens, Pierre Perret, Joe Dassin, and the Soviet singer Alla Pugachova. He is a member of various associations, including the National Air and Space Academy.

Although Chrétien is a French Air Force officer, he is on detachment to CNES for his spaceflight activities. His first detachment began in 1980, and this was renewed in late 1984 for a further five years. He will remain available for future flights with the Shuttle (NASA), Salyut (Soviet Union) or Spacelab (ESA).

Chrétien and Patrick Baudry have the distinction of being the only spacemen from outside the USA and Soviet Union to have trained on both countries' spacecraft.

Jean-Loup Chrétien was the 10th Intercosmos cosmonaut.

Lieutenant Dumitru Dediu

NO PHOTOGRAPH AVAILABLE

Dumitru Dediu was born in 1942 in Romania.

Following the agreement to fly non-Soviet cosmonauts from member countries of the Intercosmos programme onboard Soyuz and Salyut spacecraft, Romania began a cosmonaut selection programme. The two finalists were Dediu and Dumitru Prunariu who began their training at the Yuri Gagarin Cosmonaut Training Centre in March 1978. Prunariu eventually flew the mission as research-cosmonaut onboard Soyuz 40, launched on 14 May 1981, while Dediu served as back-up, and probably as CapCom in FCC.

According to unofficial information, Dediu was in fact, chosen by the Soviet Union to be the first Romanian in space. He was named as such in a United Press report issued in Moscow shortly before the Soyuz 40 mission began. It therefore seems that Valery Bykovsky and Dediu should have flown, but when Bykovsky became disqualified, the back-up crew of Popov and Prunariu took the prime slot. Dediu was moved to back-up, and Yuri Romanenko became the new back-up commander. This switch of crews was also reported by United Press.

Very few biographical details have been released for Dediu, but it is known that he is married, and his wife's name is Veronica.

Colonel-Engineer Lev Stepanovich Demin

Lev Demin was born on 11 January 1926, in Moscow, and began work in 1942 as a turner at a drilling machine plant. In 1945, he joined the Soviet Army, and later graduated from an Air Force Communications school. As a trainee pilot, he was told to give up flying because of his short-sightedness. So he studied to be an engineer, and in 1949, entered the Zhukovsky Military Air Academy. He graduated in 1956, and in the same year, joined the CPSU. He also completed a post-graduate course at the Academy in 1961. From 1956 onwards, Demin worked as a senior research associate with an Air Force research establishment, and in 1963, he earned a degree as a Candidate of Technical Sciences.

He was selected to join the cosmonaut team in 1963, and reported for duty on 11 January. *It is thought that he may have trained for Salyut 2, but when that failed, he was re-assigned to Salyut 3. (see section entitled 'Salyut 2/Cosmos 557 Crewing' for more details.)*

Demin had to wait eleven years for his only spaceflight, which came as flight-engineer onboard Soyuz 15, launched on 26 August 1974. The flight was widely expected to involve a month-long stay onboard the Salyut 3 space-station. However, a failure in the docking and rendezvous equipment prevented the docking taking place, and Demin and Gennady Sarafanov were forced to return to Earth, landing in darkness after a mission lasting only 2 days and 12 minutes.

Soviet sources later stated that the cosmonauts had been testing an automatic docking system related to the development of an unmanned Soyuz designed to resupply manned Salyut stations with consumables. (i.e. the Progress spacecraft, which made its first appearance on 20 January 1978 with the launch of Progress 1 to the Salyut 6 space-station.)

Upon his return, Demin was awarded the title of Hero of the Soviet Union and also received the Order of Lenin. He is married, and his wife Zinaida Alexandrovna works in Star Town. They have a son, Sergei, and a daughter, Natalya, who is married and has a young son Vladimir – Demin's grandson. At the time of the flight, Demin set two new records – he was the first grandfather in space, and also the oldest man (48) ever to make a first spaceflight. The latter record stood until 1975 when it was broken by ASTP astronaut Deke Slayton at the age of 51.

Demin is head of the Soviet Chess Federation, and is an avid philatelist. He is head of Star Town's Philatelic Society, and in late 1978, he was elected Chairman of the Soviet National Philatelic Federation.

Lev Demin was the 32nd Soviet cosmonaut.

Lieutenant-Colonel Georgi Timofeyevich Dobrovolsky (1928-1971)

Georgi Dobrovolsky was born on 1 June 1928, in Odessa. During World War Two, he was captured by the Nazis and sentenced to 25 years hard labour for the possession of firearms, but fortunately managed to escape from his captors. After the War ended, he wanted to attend a merchant marine academy, but failed to gain admission. So instead, he joined the Soviet Air Force by enrolling at the Chuguyev Air Force flying school. He joined the CPSU in 1954.

Upon graduation from the flying school four years later, he served as a fighter-pilot in various units before being selected to join the cosmonaut team in 1963. He reported for duty on 11 January. He once said 'I love any kind of flying, whether it is flying an airplane, descending on a parachute, performing on a trampoline, or jumping from a parachute-training tower.'

In mid-1967, he was assigned to the Zond training group, and began preparing for a flight to the Moon. Upon the abandonment of the Soviet lunar programme, he was re-assigned to the Soyuz programme, and his first spaceflight came as commander of Soyuz 11, launched on 6 June 1971. *It is believed that Dobrovolsky was originally the back-up commander for Soyuz 10 and Soyuz 11, and was not actually scheduled to fly to Salyut 1. (see section entitled 'Salyut 1 Crewing' for more details.)*

The Soyuz 11 crew docked with the Salyut 1 space-station, and spent nearly 23 days onboard, carrying out Earth observations work and other experiments. Then Dobrovolsky, Patsayev and Volkov reboarded their Soyuz spacecraft and returned to Earth. Tragically, a pressure equalization valve opened prematurely while the descent module was still above the atmosphere, and the crew were killed following the loss of cabin oxygen and pressure. The death of the three men on 30 June 1971 was a severe blow to the Soviet manned space programme, and delayed further flights for over two years. The Soyuz 11 mission had lasted 23 days 18 hours and 22 minutes, a record which stood for nearly two years before being broken by the American Skylab 1 astronauts in 1973.

Dobrovolsky was posthumously awarded the title of Hero of the Soviet Union and also received the Order of Lenin. Following a military state funeral, his ashes were buried in the Kremlin Wall. In his honour, the Soviets named one of the space tracking vessels the

‘ Cosmonaut Georgi Dobrovolsky ’, and a street in Kaluga, the birthplace of Tsiolkovsky, was also named after him.

He was married, and he and his wife Ludmilla Timofeevna – a Maths teacher – had two daughters, Maria, born in 1959, and Natasha, born in 1961.

Georgi Dobrovolsky was the 24th Soviet cosmonaut.

Major-General Vladimir Alexandrovich Dzhanibekov

Vladimir Dzhanibekov was born on 13 May 1942 in the village of Iksander, in the Bostanlyk district of the Tashkent Region, in Uzbekistan. His father, a fireman, was killed during World War 2, and after the War, the family moved to Tashkent, the capital of Uzbekistan. His mother, a nurse, also died when Dzhanibekov was young. Upon finishing school in 1953, he entered the Suvorov Military School in Tashkent, where he took a keen interest in physics, general science and aviation. He graduated in 1959, and after 'a great deal of thought and hesitation,' he enrolled in the Physics department of Leningrad University. He became involved in the University's flying club, but soon found that he'd taken on too much – flying and studying – particularly as the journey to and from the airstrip itself took a long time.

So after only a year at the University, he left, and in August 1961, he enrolled with the V.M. Komarov Higher Military Flying School in Yeisk. He graduated in 1965 with the diploma of engineer-pilot, and for the next five years, served as a flight-instructor with the Soviet Air Force at the Yeisk school. While training a group of Indian fliers, he improved his English, and spent his evenings reading English text-books and dictionaries. He had hopes of becoming a test-pilot, but after a visit by Gherman Titov to his flying school to give a talk about the Vostok 2 spaceflight and future space programmes, Dzhanibekov decided to apply to join the cosmonaut team.

He was accepted in 1970, the same year as he joined the CPSU, and in 1973, he was assigned to the Apollo-Soyuz Test Project (ASTP). His knowledge of English was probably a decisive factor in his selection. From May 1973 until July 1975, he served as back-up to Alexei Leonov, who flew as commander of Soyuz 19, the Soviet half of the joint mission, launched on 15 July 1975. During the ASTP flight, he served as one of the Soviet CapComs, and was then re-assigned to the Salyut programme. *It is thought that he may have been in the original Soyuz 25 support crew, but stood down when Andreyev became disqualified.*

He made his first spaceflight as commander of Soyuz 27, launched on 10 January 1978. The spacecraft docked with the orbiting Salyut 6 space-station, and Dzhanibekov and Oleg Makarov became the first crew to visit an already orbiting crew. During their 5 day stay onboard the space-station, the two cosmonauts carried out a series of joint experiments with the host crew of Gubarev and Grechko. They then returned to Earth after a flight lasting 5 days 22 hours and 59 minutes.

Upon his return, Dzhanibekov was awarded the title of Hero of the Soviet Union and also received the Order of Lenin.

In December 1978, he was teamed up with Bela Magyari to train for the joint Soviet-Hungarian spaceflight, and served as back-up to Valery Kubasov, the commander of the joint mission, Soyuz 36, launched on 26 May 1980.

His second spaceflight came as commander of the joint Soviet-Mongolian spaceflight, Soyuz 39, launched on 22 March 1981. Dzhanibekov and Jugderdemidiin Gurragcha linked up with Salyut 6, and joined the crew of Kovalyonok and Savinykh already onboard. The two men spent nearly seven days onboard carrying out a programme of joint Soviet-Mongolian experiments, before returning to Earth after a flight lasting 7 days 20 hours and 43 minutes.

Dzhanibekov was subsequently awarded the title of Hero of the Soviet Union and also received the Order of Lenin, both for the second time. In addition, he was awarded the title of Hero of the People's Republic of Mongolia, and also received the Order of Suhbaatar and Gold Star Medal.

His third spaceflight came as commander of the joint Soviet-French spaceflight, Soyuz T-6, launched on 24 June 1982. Yuri Malyshev had originally been selected as commander, but in February 1982, it was revealed that he had become ill quite suddenly during training (in fact he was removed because of personal incompatibility according to the French) and had been replaced by Dzhanibekov. 'He has already made two spaceflights,' said Georgi Beregovoi, 'and this together with the fact that he has already struck up a friendship with the Frenchmen, meant that he was the ideal replacement. As he is a good mixer and a kindly person, he was able to join the crew without any problems.' At the time, Dzhanibekov was in training for a long-duration mission with Alexander Alexandrov onboard Salyut 7.

Dzhanibekov, Alexander Ivanchenkov and Jean-Loup Chrétien linked up with Salyut 7, and joined the crew of Berezovoi and Lebedev already onboard. During the final approach, an onboard computer failed, and Dzhanibekov had to take over and perform a manual docking. The three men spent nearly seven days carrying out a programme of joint Soviet-French experiments, before returning to Earth after a flight lasting 7 days 21 hours and 51 minutes. Upon his return, Dzhanibekov was awarded his third Order of Lenin, and the French Order of the Legion of Honour.

His fourth spaceflight came as commander of Soyuz T-12, launched on 17 July 1984. Dzhanibekov, Svetlana Savitskaya and Igor Volk linked up with Salyut 7, and joined the crew of Kizim, Solovyov and Atkov already onboard.

It is possible that Dzhanibekov and Volk were also the support crew on Soyuz T-10, launched on 8 February 1984, and were training for a long-duration flight. They were then apparently sidetracked by the need to fly a repair mission following a main fuel line failure on the Salyut. (see section entitled 'Salyut 7 Crewing' for more details.)

Dzhanibekov had undergone special training in the water tank at the Yuri Gagarin Cosmonaut Training Centre to learn how to repair the faulty fuel line. He practised with the tools and procedures necessary to compress and seal the ruptured fuel line and it was originally intended that he would carry out the repair himself. However, Kizim and Solovyov, having carried out the four preparatory EVA's, asked to complete the job themselves. Therefore, Dzhanibekov delivered a section of pipeline and a video film of his water tank repair procedure, and showed how the repairs should be carried out.

He and Savitskaya instead carried out an EVA lasting 3 hours and 35 minutes during which the two tested a new electron beam device. Savitskaya used it to carry out a series of operations involving the cutting, welding and soldering of metal plates and the spraying on of a coating. During the experiments, Dzhanibekov carried out cine-photography and TV reporting. The two cosmonauts then changed places and the cycle of technological operations on other samples was repeated by Dzhanibekov. Following the successful EVA, the Soyuz T-12 crew conducted a series of experiments onboard the Salyut before returning to Earth after a flight lasting 11 days 19 hours and 14 minutes.

Upon his return, Dzhanibekov was awarded his fourth Order of Lenin.

Upon the completion of the Soyuz T-12 mission, the original flight schedule was resumed, and Viktor Savinykh should then have flown on Soyuz T-13 with Vladimir Vasyutin and Alexander Volkov. *(see section entitled 'Salyut 7 Crewing' for more details.)*

However, when Salyut 7 suffered an onboard failure which left it without power, another repair mission became necessary. As a manual docking was required, Dzhanibekov was given the assignment, following his experience on Soyuz T-6. Savinykh was paired with him at the end of March 1985, as extensive onboard repairs were anticipated, and an experienced flight-engineer was needed.

The two men were launched onboard Soyuz T-13 on 6 June 1985, and in making the flight, Dzhanibekov became the first cosmonaut to fly in space five times. The two men linked up with Salyut 7 using a new manual docking system in a bid to rescue the lifeless station. They found the interior to be dark, freezing cold, and encrusted in ice. Working out of the Soyuz, which provided their only source of warmth, the two men carried out a series of repairs to the Salyut's electrical system, eventually returning it to a fully operational state. Their repair work salvaged the station, and enabled the Soviets to resume the planned flights to Salyut 7.

On 2 August, the two men carried out an EVA lasting 5 hours, during which they tested an upgraded spacesuit and installed an additional solar panel. They also installed a Medusa experiment and a device to collect dust from Halley's Comet.

Once the Salyut was declared operational, a new crew was launched to the station onboard Soyuz T-14. The two crews worked together for a week, and then Dzhanibekov returned to Earth with the Soyuz T-14 flight-engineer, Georgi Grechko. In doing so, he became part of the first partial replacement of a crew. Savinykh remained onboard the Salyut with the new arrivals, Vasyutin and A.Volkov, thus restoring the original Soyuz T-13 crew.

Dzhanibekov had spent 112 days 3 hours and 12 minutes in space, and upon his return, he was awarded his fifth Order of Lenin, and was promoted to Major-General.

He is married, and his wife Lilia Munirovna is a school singing tutor, a graduate of the Conservatoire, who now works at Star Town. They have two daughters, Olga, born in 1974 or 1975, and Inna, born in 1969. Dzhanibekov has been described as soft-spoken and calm tempered, never being prone to rash decisions. Shatalov once described him as 'modest ... a man of few words' and Klimuk said that he 'never hurries, doing everything very thoroughly. He likes engineering, and attends training sessions in simulators with pleasure.'

Dzhanibekov is a ham-radio operator, and he built his own hi-fi system. Shatalov reported that 'when he was training under the Soyuz-Apollo programme, he surprised us by his passion for radio-equipment.' He reads extensively on radio-engineering and

Shakespeare's sonnets in English, many of which he can recite from memory. Like Alexei Leonov, he is a good artist, and his cartoons are a permanent feature on the cosmonauts' walls. He has designed a set of stamps, issued in September 1980, and with Herman Komlev, co-designed a set issued in March 1981. When asked about his hobbies, he once replied 'Actually, I suppose I could say that my work's my hobby – there's nothing else that I'm so keen on.'

Known to his friends as Volodya, Dzhanibekov is Vice-President of the Soviet-Spanish Friendship Society, and is a Deputy to the Uzbek SSR Supreme Soviet. He has received many awards and honours apart from those already listed. They include:- The 20th Victory Anniversary of the Great Patriotic War; The 50th Anniversary of the Soviet Armed Forces, for irreproachable service, III Grade; and a Gold Medal from the Mayor of Lyons, presented in June 1978, in recognition of the Soviet Union's role in the exploration of space.

Vladimir Dzhanibekov was the 43rd Soviet cosmonaut, and he has spent a total of 145 days 15 hours and 59 minutes in space.

Mohammad Al Fares

Mohammad Al Fares is one of the two Syrian cosmonauts chosen for the joint Soviet-Syrian spaceflight.

He is a pilot in the Syrian Air Force, and began training at the Yuri Gagarin Cosmonaut Training Centre in late 1985.

At the time of his selection he was aged 34. He is married, and he and his wife Gind Akil have a son Kutaib, born in 1980 or 1981, and a daughter Gadil, born in 1979 or 1980. No other biographical details have been released.

Lieutenant-Colonel Bertalan Farkas

Bertalan Farkas was born on 2 August 1949, in the village of Gyulahaza in North-East Hungary, the son of Laiosh and Erzhebet Farkas. His father was a shoe-maker and worked in a shoe-maker's co-operative workshop in Kisvarda, where Farkas attended the Gyorgy Bessenye Secondary School. He was interested in aeronautics from a very early age, and while a youth, joined the flying club of the Hungarian National Defence Association at Nyiregyhaza, and began making glider flights. After graduating from secondary school in 1967, he enrolled in the Gyorgy Kilian Aeronautical College in Szolnok. Upon graduation in 1969, he went to the Soviet Union to continue his training.

Farkas was commissioned as a Second Lieutenant in the Hungarian People's Army in 1972, and joined the Air Force. He joined the Hungarian Socialist Workers Party in 1976, and in 1977, he qualified as a fighter-pilot first class. In 1978, he was promoted to Captain (Squadron Leader), having been especially successful in complicated manoeuvres at high altitude, and in manoeuvring during combat training. He was decorated on five occasions, and repeatedly praised for his performance of his duties.

Following the agreement to fly non-Soviet cosmonauts from member countries of the Intercosmos programme onboard Soyuz and Salyut spacecraft, Hungary began a cosmonaut selection programme. The two finalists were Farkas and Bela Magyari and they were sent to begin their training at the Yuri Gagarin Cosmonaut Training Centre, arriving on 20 March 1978. Following nine months of general training, Farkas was teamed up with Valery Kubasov in December 1978, and flew as research-cosmonaut onboard Soyuz 36, launched on 26 May 1980.

The two men linked up with Salyut 6, and joined the crew of Popov and Ryumin already onboard. Kubasov and Farkas spent nearly seven days onboard carrying out a programme of joint Soviet-Hungarian experiments. They then returned to Earth after a flight lasting 7 days 20 hours and 46 minutes. During the flight, Farkas lost 3 kg.

Following the successful completion of the mission, he was promoted from Captain to Lieutenant-Colonel and received many awards and honours including:- The title of Hero of the Soviet Union and Order of Lenin; the title of Hero of the Hungarian People's Republic; the Golden Wreath badge of the Hungarian-Soviet Friendship Society; and the title of Cosmonaut of the Hungarian People's Republic.

In June 1985, he and Vladimir Kovalyonok took part in the ground studies at Kursk Oblast as part of the Kursk-85 experiment conducted by the Soyuz T-13 cosmonauts.

Farkas is married, and his wife Anniko is a photographer. They have a son, Gabor, born in 1971, and a daughter, Aida, born in 1977. During his spaceflight, Farkas demonstrated part of his character by taking his daughter's favourite toy rabbit along with him. When asked why, he replied that it was 'a reminder of what I treasure most in life.'

He has been described as a 'highly disciplined and politically thoroughly trained officer, with a high degree of self-control.' Kubasov said he was 'very serious, a good specialist, with very quick reactions – grasps everything at once, a charming person and sociable – I value his sincerity and sense of responsiblity. You can always rely on him.'

Farkas is known as Bertsi to his friends – an abbreviation of Bertalan. When asked what his hobbies were, he replied that he was very keen on football – so much so that he had been at a loss at what to choose for a career – football or flying.

Bertalan Farkas was the 5th Intercosmos cosmonaut.

Dr. Konstantin Petrovich Feoktistov

Konstantin Feoktistov was born on 7 February 1926, in Voronezh. At the age of 10, he worked out a detailed plan to explore the Moon, which involved studying mathematics and physics in order to understand the Tsiolkovsky formulae. During World War 2, he served as an army scout, and in 1942, he was captured by the Nazis, who put him before a firing squad. By a miracle, he was not fatally wounded, and when night fell, he crawled out from under the bodies and managed to reach Soviet troops. He made a speedy recovery, and in 1943, he entered the E.N. Bauman Moscow Higher Technical School, graduating in 1949 with a Candidate of Technical Sciences degree.

He worked as an industrial engineer, then joined the space programme in 1955, as a member of the Bushuyev team at Sergei Korolyov's design bureau, and also gave lectures on spacecraft construction to the first cosmonauts. He was said to be one of the leading Vostok designers as well as one of those working on the Voskhod. In particular, he headed the design team in charge of developing the Vostok/Voskhod landing system.

It is believed that he volunteered to fly onboard one of the Voskhods, and as a result, was officially selected to join the cosmonaut team in 1964. His first spaceflight came as researcher onboard Voskhod 1, launched on 12 October 1964, and during the mission, Feoktistov tested the new equipment and systems onboard the spacecraft, as well as carrying out geophysical and astronomical observations. Then together with Boris Yegorov and Vladimir Komarov, he returned to Earth after a flight lasting just 1 day and 17 minutes.

Upon his return, Feoktistov was awarded the title of Hero of the Soviet Union and also received the Order of Lenin. He then appeared to give up his temporary role of cosmonaut to return to leading the spacecraft design team. He is one of the chief designers of the Salyut series of space-stations, and is still actively involved in cosmonaut training. After Voskhod 1, he wrote a thesis on spaceflight which earned him a DSc. degree in 1966. In May 1975, he served as flight-director for the Soyuz 18/Salyut 4 mission, taking over from Alexei Yeliseyev during his ASTP duties.

In 1982, a Soviet book appeared which contained an interview with Feoktistov, in his capacity as 'a member of the crew of the forthcoming Soyuz T-3, a spacecraft which he had helped construct.'

He explained why he intended to make a second spaceflight. 'If you remember, in 1964, on the Voskhod my duties were those of flight-engineer. The Chief Designer, Sergei Korolyov, gave me the job, as one of those involved in its design, of testing the Voskhod in flight, together with Vladimir Komarov and Boris Yegorov. Actually working on the Voskhod in space provided us with much material for modernising the craft and then later providing the multi-purpose Soyuz, and designing the orbital Salyut stations.

I think it's high time I went up again to take a look for myself at how the systems of the new Soyuz T function, as well as those of the Salyut, which has surpassed all expectations in operating so long. An exhaustive and detailed analysis of the systems operation in space will do me a lot of good.'

In the event, Soyuz T-3 was launched on 27 November 1980 carrying a crew of Leonid Kizim, Oleg Makarov and Gennady Strekalov. However, photographs have since appeared which show Kizim, Makarov and Feoktistov training for a sea-recovery, and Feoktistov training in a Soyuz-T simulator. It appears therefore that he had to be dropped from the mission at a very late stage, presumably for health reasons, and was replaced by Strekalov. Had he made the flight, he would have been nearly 55.

He is married, and he and his wife Galina Nikolaevna have four children, including a son, Andrei, born in 1962, a daughter, Galina, and a son, Kostya, born in 1982 or 1983.

It is notable that Feoktistov has never joined the CPSU. He is the author of a number of books, including 'Our Engineers' published in 1982; 'About Spaceships' also published in 1982; and '7 Steps into the Sky' published in 1984.

Konstantin Feoktistov was the 8th Soviet cosmonaut.

Major-General Anatoli Vasilyevich Filipchenko

Anatoli Filipchenko was born on 26 February 1928, in the village of Davydovka, in the Voronezh Region. He went to school for seven years in Ostrogozhsk, before leaving in 1942 to help his mother at a motor repair plant while his father was away during the War. At the age of 15, he began work as a lathe-operator.

In 1947, he enrolled in the Chuguyev Air Force School and graduated with honours in 1950. He joined the Soviet Air Force and rose from junior pilot to senior pilot instructor, and later, to Pilot 1st Class. He joined the CPSU in 1952, and enrolled in a correspondence course with the Red Banner (now Gagarin) Air Force Academy in 1955, graduating in 1961.

He was selected to join the cosmonaut team in 1963, and reported to Star Town on 11 January. Following his selection, he attained the rank of test-pilot, 3rd class, having flown approximately 20 different types of plane during his career. During his cosmonaut training, he continued to fly supersonic aircraft.

He was involved in some way with the flight of Voskhod 2, launched on 18 March 1965, probably as a member of the support crew.

His next assignment was as back-up to Vladimir Shatalov, the commander of Soyuz 4, launched on 14 January 1969. He made his first spaceflight as commander of Soyuz 7, launched on 12 October 1969. Filipchenko, Vladislav Volkov and Viktor Gorbatko carried out rendezvous manoeuvres and other experiments in a joint flight with the Soyuz 6 and 8 spacecraft, before returning to Earth after a flight lasting 4 days 22 hours and 40 minutes.

Upon his return, Filipchenko was awarded the title of Hero of the Soviet Union and also received the Order of Lenin.

According to one Western report, a heavy Zond-Proton countdown was conducted on 3 and 4 July 1969, and the cosmonaut crew present at Baikonur was thought to consist of Filipchenko, Kubasov and Shonin. Their launch was cancelled when the G-vehicle failed shortly after lift-off. (see section entitled 'The Zond programme' for more details.)

His next official assignment was as back-up to Andrian Nikolayev, the commander of Soyuz 9, launched on 1 June 1970. However, Vasily Lazarev has also been officially named as the back-up, and it is believed that Filipchenko was actually the commander of the support crew.

It is thought that Filipchenko may have served as a back-up crewmember for the Cosmos 557 Salyut attempt in May 1973. (See section entitled 'Salyut 2 / Cosmos 557 Crewing' for more details.) His next assignment also came in 1973, when he was named as the commander of the reserve Soyuz for the Soviet half of the Apollo-Soyuz Test Project. From May 1973 he trained for the mission in both the United States and the Soviet Union. As part of the preparations for the joint flight, Filipchenko made a second spaceflight as commander of the Soyuz 16 ASTP rehearsal mission, launched on 2 December 1974. He and Nikolai Rukavishnikov carried out a dress rehearsal of the ASTP Soyuz flight programme, including important docking tests, before returning to Earth after a flight lasting 5 days 22 hours and 24 minutes.

Filipchenko was subsequently awarded the title of Hero of the Soviet Union and also received the Order of Lenin, both for the second time.

He has received many other awards and honours including:- the Valour Award; the Order of the Banner with diamonds (from Hungary); the Order of the Star of the Republic, 5th degree (from Romania); a number of Soviet medals and medals from other countries; the FAI Gagarin medal; the Tsiolkovsky Gold Medal of the Soviet Academy of Sciences; and the Small Gold Medal of the ENEA.

In addition, he has been made an honorary citizen of Karaganda, Kaluga, Sumy, Chita, Lipetsk, Ostrogozhsk, Davydovka and Arkalyk.

In September 1976, he was said to be closely involved with the Soyuz 22 flight, and served as Chief CapCom. In 1978, he served as CapCom on the Soyuz 31 mission.

In early 1979, he was elected Chairman of the newly founded Federation of Cosmonauts, and served in this capacity until 1983. He is also Vice-Chairman of the Central Board of the Soviet-GDR Friendship Society.

He is married, and he and his wife, Elizaveta Alexandrovna – a housewife – have two sons, Alexander and Igor. Filipchenko is known to his friends as "Filip", and has been described as a strong-willed, but modest person, who is friendly, personable, and easy going. He is said to be an ardent sportsman, a sharp-shooter and hunter, and enjoys underwater speargun fishing, as well as touring in his car. He has written an autobiography entitled 'Safe Orbits' or 'Reliable Orbits', published in 1978.

Anatoli Filipchenko was the 19th Soviet cosmonaut, and he has spent a total of 10 days 21 hours and 4 minutes in space.

Colonel Yuri Alexeievich Gagarin (1934-1968)

Yuri Gagarin was born on 9 March 1934, in the village of Klushino, some 100 miles (161 km) West of Moscow, situated near the town of Gzhatsk (now named Gagarin in his honour) in the Smolensk Region. His father, Alexei Ivanovich, was a collective farmer, and his mother, Anna Timofeyevna, was a dairymaid. He started school in 1941, but his schooling was interrupted by the Nazi invasion. The Gagarin's house was occupied, and the family had to live in a dug-out. When the Germans retreated, they took two of Yuri's sisters, but they were released after the War.

The Gagarin family moved to Gzhatsk after the War, where Yuri re-commenced school. (The name Gagarin incidentally, derives from the Soviet word for wild duck!). His favourite subject was arithmetic, and he also developed a keenness for science.

Upon finishing school in 1950, he enrolled in a foundryman's school at Lyubertsky, near Moscow, and also attended evening classes. Students at the school combined factory work and study while they prepared for jobs in Soviet industry. After a year however, Gagarin was accepted as a student at a four year technical school where he could finish his secondary education while he also studied foundry work. His new school was in Saratov on the Volga river, and had a good flying school and an airfield nearby.

As soon as he was eligible, during his fourth year at Saratov, he enrolled at the flying school, the Saratov Aero Club, where he studied the theory of flight and other aviation subjects at night. His course included a parachute jump, which he made from the wing of a traininig plane. After his successful jump, an instructor gave him a ride in a Yak-18 fighter plane. 'That first flight ' he later recalled 'filled me with pride and gave meaning to my whole life.'

In the Spring of 1955, he graduated with honours as a foundryman-technician, and also received his ground school diploma from the flying school. He was determined to become a pilot, and instead of going to work as a foundryman, he spent the summer at an aviation camp, where he learned to fly the Yak-18. At the end of the Summer, he became an aviation cadet at the Orenburg Pilot Training School, graduating two years later as a pilot, first-class. He then joined the Soviet Air Force and served as a fighter-pilot.

His first posting was at Orenburg, where he met his future wife, Valya (Valentina), a nursing student. She apparently made regular visits to watch him marching, but couldn't always see him, as he was not allowed to march in the front ranks because he was too short! His lack of height was to cause him further problems when he was assigned to flying MiG fighters, and had to resort to using a cushion to reach the controls!

Gagarin was promoted to Lieutenant in late 1957, and married Valya that November. Soon afterwards, he volunteered for a tour of duty in the Arctic, while Valya completed nursing school in Moscow. After graduating, she joined him for the Arctic winter. The following spring, she gave birth to their first child, a 7lb 7oz (3.4 kg) daughter, Lena. (Lena, or Yelena is the diminutive of Lenochka.)

In late 1959, it is said that Lieutenant Gagarin applied for admission to the astronaut training group 'if such a group exists'. He was called up, along with several dozen other candidates before a strict medical board. He passed all the tests, and on his 26th birthday i.e. 9 March 1960, he was transferred to the cosmonaut team. The work was so secret that he wasn't even allowed to tell Valya. He joined the CPSU in the Summer of 1960.

On 7 March 1961, his second daughter, was born. 'It's Spring, so let's give her a Spring name – Galya ' said Gagarin. (Galya is the diminutive of Galchonok – young Daw. Various reports give differing names – Galochka, Galinka and Galina.) Shortly afterwards, Gagarin told Valya that he was not only training for flights into space, but had been chosen as the first man to fly.

The flight into space – and History – began when Vostok 1 was launched on 12 April 1961. Gagarin made a single revolution of the Earth before returning after a flight lasting 1 hour and 48 minutes.

Upon his return, he was awarded the title of Hero of the Soviet Union and also received the Order of Lenin. He made many tours of the World, and was showered with honours and awards in recognition of his great feat. A crater on the lunar farside was named in his honour, and during a visit to London, he was presented with a Gold Medal from the British Interplanetary Society.

In the years that followed, he served as a tutor to the other cosmonauts, and became the official deputy to Lieutenant-General Nikolai Kamanin, the chief of the Soviet space programme commission. In 1963, he became Commander of the Cosmonaut Detachment, and served as CapCom for the Vostok 3 and 4 missions, as well as technical director for Vostok 6. In addition, he was made responsible for all projects connected with lunar exploration. He began studying at the Zhukovsky Military Air Academy in September 1961, and graduated in February 1968. He also co-wrote a book with Vladimir Lebedev entitled 'Psychology and Space.'

Valya later wrote that during holidays from the training, Gagarin 'would enjoy having people round. When listening to funny stories, he could not hold back his ringing and infectious laughter. The boys were drawn to him and good-naturedly called him "Gagara" (Loon!).'

Gagarin insisted that he would make a second, maybe third spaceflight, and this, coupled with his responsibilty for lunar projects, led many people to believe that he would lead the Soviet lunar landing team. Apparent confirmation came from Valentina Tereshkova when, during a visit to Havana, she told the Cubans 'The Moon team has already been picked. Major Gagarin is head of it, and I am on it.'

He headed the Technical Commission of the Soviet Water Skiing Federation, and was also a member of the Komsomol Central Committee, as well as head of the Soviet-Cuban Friendship Society. In addition, he served on the Council of the Union of the Supreme Soviet from 1962-66, and the Council of Nationalities of the Supreme Soviet from 1966-68. In 1967, he became the first foreigner to be awarded the title of Hero of the People's Republic of Bulgaria during a visit to that country. His wife currently works as an assistant in one of the bio-medical laboratories at the Yuri Gagarin Cosmonaut Training Centre.

In August 1966, Gagarin began taking an active part in the space programme again, when he was assigned as back-up to Vladimir Komarov, the pilot of Soyuz 1, launched on 23 April 1967. Komarov was killed during the return to Earth, and Gagarin was assigned to fly the next manned mission, Soyuz 3. Normal crewing policy dictated that he should fly the mission anyway, but of all the cosmonauts, Gagarin would have had the greatest influence in restoring confidence in the Soyuz spacecraft. As events turned out however, he was never to fly a second mission.

It appears that at some stage after the Soyuz 1 accident, Gagarin lost his flight status. According to one Western report, he was involved in a car crash which resulted in the scar seen over one eye. Alternatively, he may simply have been grounded following Vostok 1, it being considered that he was too valuable to risk. Whatever the reason, it seems that he had to regain his flight status, and he began flying under supervision in early March 1968, after passing a medical examination on 12 March.

He asked to be relieved of all other duties, and made his first flight on 13 March in a two-seater fighter. It was said that his break from flying did not seem to have affected his piloting abilities. He flew again on 19 and 20 March, and this was said to be 'a further step towards flying solo.'

On 22 March he flew again, and his trainer Lieutenant-Colonel Alexander Ustenko 'knew that Gagarin would soon be allowed to make a solo flight, and therefore checked his flying technique and landing with special care.'

On 23 March, General Kamanin, discussing Gagarin's training programme, said 'It's a very fast pace, Yuri,' evidently worried that he was overworking. However, he did not try to dissuade Gagarin from flying, and consequently from training for another spaceflight. In his opinion, he should never have stopped flying.

On 27 March 1968, at 10.19 AM, he and Colonel Vladimir Seryogin took off from the airfield adjoining Star Town for another routine flight in a two-seat MiG 15. Several minutes later, the two men requested permission to alter course, and this was granted. Nothing more was heard from them. With mounting alarm, it was decided to send out search planes, and they found the MiG's smoking wreckage approximately 30 miles (48.3 km) East of Moscow. The plane's chronometer had stopped at 10.31.

Gagarin's ashes would have been buried in the Kremlin wall, but, it was revealed in 1984, his body was never found. In his honour, the Soviets named one of the space tracking vessels the 'Cosmonaut Yuri Gagarin.'

His death was a terrible blow to the Soviet space programme, coming as it did less than a year after the death of Komarov. He had been an extremely able and popular man, deeply respected by his colleagues. They had described him as 'a man of principles, decisive, bold and steadfast.' He had 'a straightforward and frank personality, with a warm smile and quick wit.' He was also very good at putting people at their ease in his presence. Nikolai Kamanin had once said 'He was lively, had a winning personality and was a good mixer.

He had an insatiable thirst for knowledge.' Those who met him reported that he was an unassuming and pleasant man who seemed quite unaffected by the publicity that his exploit had brought him.

Yuri Gagarin's contribution to manned spaceflight was great indeed. He flew into the unknowns of Space, where no man had been before and in doing so, he opened the way to the Stars. His name will live forever in the annals of manned spaceflight.

Captain Majdarzavyn Ganzorig

NO PHOTOGRAPH AVAILABLE

Majdarzavyn Ganzorig was born on 6 February 1949, in the village of Ich-Chuzirt, Cercerleg Settlement, Ara-Changay Province, Mongolia, the son of a shepherd.

He graduated from basic and secondary schools in the town of Cercerleg, and in 1969, enrolled in the Kiev Technical Institute, where he specialized on the problems of thermal energy. He graduated from the Institute in February 1975, and started work as a specialist on controlling and measuring apparatuses and automation in Ulan-Bator, the capital of Mongolia.

In 1977, he was approved as a scientist at the Physical and Technical Institute of Mongolia, and he worked on the automation of scientific experiments. In 1978, he was awarded the rank of Captain in the Mongolian Air Force, part of the Mongolian Army. He joined the Mongolian People's Revolutionary Party in 1979.

Following the agreement to fly non-Soviet cosmonauts from member countries of the Intercosmos programme onboard Soyuz and Salyut spacecraft, Mongolia began a cosmonaut selection programme. The two finalists were Ganzorig and Jugderdemidiin Gurragcha, who began their training at the Yuri Gagarin Cosmonaut Training Centre in April 1978. Ganzorig teamed up with Vladimir Lyakhov, and Jugderdemidiin Gurragcha eventually flew the mission as research-cosmonaut onboard Soyuz 39, launched on 22 March 1981, while Ganzorig served as back-up, and probably as CapCom in FCC.

He is married, and he and his wife – a school-teacher – have two sons.

Colonel-Engineer Yuri Nikolaievich Glazkov

Yuri Glazkov was born on 2 October 1939, in Moscow. He enrolled at the Air Force Higher Engineering College in Kharkov and graduated in 1962, following which he began work as an engineer in the Soviet Air Force.

He was selected to join the cosmonaut team in October/November 1965, but had to wait 12 years before making his first spaceflight. He joined the CPSU in 1966, and became an expert pilot as well as a qualified parachute instructor. He also became a specialist in spacecraft control systems, and took part in the ground control of several Soyuz flights, including the joint Soyuz 6, 7 and 8 mission in 1969.

In 1974, he served as a CapCom onboard one of the Soviet tracking ships during the Salyut 3 missions, and in late 1974, he began training for a Salyut mission with Viktor Gorbatko, whilst continuing his studies for a degree. He wrote a thesis on EVA, and was awarded a Candidate of Technical Science degree in 1974. In the early part of 1975, he was onboard the tracking ship 'Cosmonaut Yuri Gagarin', acting as CapCom for the Soyuz 17 and 18 missions.

It is believed that his first assignment was as flight-engineer on the support crew for Soyuz 21, launched on 6 July 1976. (see section entitled 'Salyut 5 Crewing' for more details.) He then served as back-up to Valery Rozhdestvensky, the flight-engineer onboard Soyuz 23, launched on 14 October 1976. Following the failure of that mission, Glazkov and Viktor Gorbatko were launched into space onboard Soyuz 24 on 7 February 1977. Gorbatko said at the time 'According to the documents, the crew was formed two and a half years ago, but actually much earlier. Some five years ago training schedules placed us in the same training group.' This would seem to indicate that Glazkov was assigned to a group of cosmonauts which began training for military Salyut missions in 1972.

The two men docked with the Salyut 5 space-station and spent over two weeks onboard, carrying out medical and technical experiments together with Earth observations. They returned to Earth after a flight lasting 17 days 17 hours and 26 minutes.

Upon his return, Glazkov was awarded the title of Hero of the Soviet Union and also received the Order of Lenin.

In late 1977 and early 1978 he served as CapCom for the Soyuz 26 and 27 missions, and in late 1980, he helped to train the French cosmonauts upon their arrival at Star Town. In 1982, it was reported that he saw the Soyuz T-5 crew off on their mission, and was also one of the first to greet them upon their return. In late 1983, following their return to Earth, he accompanied the Soyuz T-9 crew on their flight back to the cosmodrome, and in 1984, he was involved in the training of the Soyuz T-12 crew.

He is married, and is said to be a quiet, modest man, with a reputation for hard work and an excellent dry sense of humour. He is Vice-President of the Soviet-People's Republic of Southern Yemen Friendship Society, and is Chairman of the Soviet Badminton Federation. He is also the co-author of a book entitled 'A Spaceship in Orbit', published in 1984.

Yuri Glazkov was the 39th Soviet cosmonaut.

Major-General Viktor Vasilyevich Gorbatko

Viktor Gorbatko was born on 3 December 1934, in the village of Ventsy-Zarya in the Caucasus region of the Krasnodar Territory in Southern Russia, and spent his childhood in the Kuban. Upon finishing school, he was determined to fly, so he went to a military flying school and then, in 1953 to the Bataisk Military Aviation College near Rostov, from which he graduated in 1956. He then spent a number of years in the Soviet Air Force, and at one stage served in the same regiment as Yevgeny Khrunov. Gorbatko joined the CPSU in 1959.

He was selected to join the cosmonaut team in 1960, and his first assignment was as back-up to Pavel Belyayev, the commander of Voskhod 2, launched on 18 March 1965. According to one US source, Gorbatko was relieved of his Voskhod 2 back-up duties for medical reasons, and this appears to have been confirmed in Popovich's book 'Test By Space and Earth' published in 1982. The book conveys that Gorbatko was originally back-up commander for Voskhod 2 but was injured or became ill, and was replaced by the cosmonaut known as Dmitri. He was then made 2nd back-up to Yevgeny Khrunov (i.e. on the support crew.)

While continuing his training, Gorbatko studied at the Zhukovsky Military Air Academy, and he graduated in 1968.

On 23 April 1967, Vladimir Komarov was launched onboard Soyuz 1. He was brought back to Earth after just over a day in orbit, and was killed when the spacecraft's parachute tangled following re-entry. It is now clear, despite official Soviet denials at the time, that a second spacecraft, Soyuz 2, was to have been launched on 24 April. Following docking, an exchange of crews would have taken place.

The Soyuz 2 prime and back-up crews have since been identified by Western observers. The prime crew were Valery Bykovsky, Alexei Yeliseyev and Yevgeny Khrunov, and the back-up crew were Pavel Belyayev/Andrian Nikolayev, Valery Kubasov and Viktor Gorbatko. Gorbatko's next assignment therefore was as back-up to Yevgeny Khrunov, who would have made an EVA to transfer from Soyuz 2 to Soyuz 1. However, following the initial problems with Soyuz 1, the Soyuz 2 launch was cancelled.

An unmanned Soyuz 2 subsequently flew in a rendezvous operation with Soyuz 3, piloted by Georgi Beregovoi, and in January 1969, nearly two years late, the mission intended for Soyuz 1/2 was carried out by the Soyuz 4/5 spacecraft. The prime and back-up crews of Soyuz 2 were given the same assignments on Soyuz 5 – only the commanders had changed. Gorbatko therefore again served as back-up to Yevgeny Khrunov, test-engineer onboard Soyuz 5, launched on 15 January 1969.

His first spaceflight came as test-engineer onboard Soyuz 7, launched on 12 October 1969. Soyuz 7 carried out rendezvous manoeuvres and other experiments in a joint flight with the Soyuz 6 and 8 spacecraft, before returning to Earth after a flight lasting 4 days 22 hours and 40 minutes.

Upon his return, Gorbatko was awarded the title of Hero of the Soviet Union and also received the Order of Lenin.

According to one Western report, Gorbatko broke his ankle in a parachute jump during training for the Soyuz 7 mission. It was slow in healing, and therefore he began to fall behind the crew training schedule. It seemed that he might lose his place on the mission to his back-up, Georgi Grechko, but, according to the report Gorbatko 'abrasively manoeuvred to assure himself of the seat.'

However, according to a second Western report which gave its sources as participants in the Zond programme, Grechko was the prime and was only replaced by Gorbatko when he fell ill with influenza shortly before the mission.

Gorbatko spent the next few years taking part in the preparations for a number of Soyuz missions, acquiring a reputation as an expert ground-controller. He continued to fly, and in addition, learnt to fly helicopters.

In July 1974, he served as CapCom on the Soyuz 14/Salyut 3 mission, and in late 1974, he began training for a Salyut mission with Yuri Glazkov. *It is believed that his next assignment was as commander of the support crew for Soyuz 21, launched on 6 July 1976. (see section entitled 'Salyut 5 Crewing' for more details.)* He then served as back-up to Vyacheslav Zudov, the commander of Soyuz 23, launched on 14 October 1976. He also acted as a CapCom during the mission.

Following the failure of Soyuz 23, Gorbatko and Glazkov were launched into space onboard Soyuz 24 on 7 February 1977. Gorbatko said at the time 'According to the documents, the crew was formed two and a half years ago, but actually much earlier. Some five years ago training schedules placed us in the same training group.' This would indicate that Gorbatko was assigned to a group of cosmonauts which began training for military Salyut missions in 1972.

The two men docked with the Salyut 5 space-station and spent over two weeks onboard, carrying out medical and technical experiments together with Earth observations. They returned to Earth after a flight lasting 17 days 17 hours and 26 minutes.

Gorbatko was subsequently awarded the title of Hero of the Soviet Union and also received the Order of Lenin, both for the second time.

He served as CapCom for the Soyuz 30 mission in June 1978, and his next assignment was as back-up to Valery Bykovsky, the commander of Soyuz 31, launched on 26 August 1978. *There is some evidence to suggest that Georgi Grechko was originally in training for the mission. He was moved into the Soyuz 26 crew following the failure of Soyuz 25, and his place on the Soyuz 31 mission was then taken by Gorbatko.*

In September 1979, Gorbatko was teamed up with Pham Tuan, one of the two Vietnamese cosmonauts, and they were launched into space onboard Soyuz 37 on 23 July 1980. The two men docked their spacecraft with the orbiting Salyut 6 space-station and joined the crew of Popov and Ryumin already onboard. They then carried out a series of joint Soviet-Vietnamese experiments and observations before returning to Earth after a flight lasting 7 days 20 hours and 42 minutes.

Upon his return, Gorbatko received many awards, including the title of Hero of Socialist Labour of Vietnam, the Order of Ho Chi Minh, 1st Class, and the Order of Lenin for the third time. He was also made an honorary citizen of Dzhezhazgan, the nearest town to the landing site.

He was then put in charge of the training programme for the joint Soviet-Mongolian spaceflight, Soyuz 39, launched on 22 March 1981.

Gorbatko is married, and he and his wife, Valentina – a doctor – have two children, Irina and Marina. He takes an active part in public life, and is President of the Soviet-Mongolian Friendship Society, and Chairman of the Soviet Federation of Parachute Sport. He holds the title of Hero of the Mongolian Peoples Republic.

Viktor Gorbatko was the 21st Soviet cosmonaut, and he has spent a total of 30 days 12 hours and 48 minutes in space.

Georgi Mikhailovich Grechko

Georgi Grechko was born on 25 May 1931, in Leningrad. At the outbreak of World War II, he was evacuated to Chernigov, in the Ukraine, and did not return to Leningrad until 1944. He was very interested in mathematics and physics at school, and later enrolled at the Leningrad Institute of Mechanics, graduating in 1955 with honours, his speciality being computers. He then began work in the design bureau headed by the late Sergei Koroloyov. Grechko was involved in the computer work which calculated the trajectories for the first Soviet space shots in 1957. He joined the CPSU in 1960.

At the design bureau, he showed himself to be an able engineer, with plenty of initiative. In 1967, he earned a Candidate of Technical Sciences degree on the basis of practical work connected with the problems of landing automatic stations on the Moon. Grechko is clearly one of the Soviet Union's leading spacecraft designers, and his work on the Luna soft-landers has been widely acknowledged in Soviet literature. He speaks English quite fluently.

He was interviewed together with Feoktistov in 1964, but was not officially selected to join the cosmonaut team until 1966. He arrived in Star Town on 5 September 1966, and was described as a mild and even-tempered man with a high standard of theoretical training and rich engineering experience.

In his diary, Vladislav Volkov wrote of a cosmonaut engineer selected in 1966, and identified only as 'Georgi', who severely injured his back in a parachute jump in the summer of 1966. Following the accident, he was removed from the programme, although Volkov did not say whether Georgi was permanently crippled or not. It would seem likely that this cosmonaut was in fact Georgi Grechko, and this has been confirmed by one Western report which stated that he actually broke his back, taking more than a year to recover.

In mid-1967, Grechko was assigned to the Zond training group, and began preparing for a flight to the Moon. *According to one report, he was the chief electronics instrumentation engineer for the Soviet Lunar Module, and was scheduled to fly on the second lunar landing mission. (See section entitled 'The Zond programme' for more details.) His experience with the unmanned Luna flights would certainly have been of great use in the manned programme.*

Upon the abandonment of the Soviet lunar programme, he was re-assigned to the Soyuz programme, and his first assignment was as back-up to Viktor Gorbatko, the test-engineer onboard Soyuz 7, launched on 12 October 1969. *According to one report, he was originally on the prime crew for the mission, but fell ill with influenza shortly before launch and was replaced by his back-up, Gorbatko.* Grechko then served as back-up to Vitaly Sevastyanov, the flight-engineer onboard Soyuz 9, launched on 1 June 1970. At the end of 1971, he was paired with Alexei Gubarev. *It is thought that the two men may have served as the support crew for the Cosmos 557 Salyut launched in May 1973. (See section entitled 'Salyut 2/Cosmos 557 Crewing' for more details.)* Grechko's next assignment was yet another back-up duty, this time to Oleg Makarov, the flight-engineer onboard Soyuz 12, launched on 27 September 1973.

His first spaceflight came as flight-engineer onboard Soyuz 17, launched on 10 January 1975. Grechko and Gubarev docked with the Salyut 4 space-station and spent nearly 30 days onboard, carrying out Earth-resources work and astronomical observations. The two men returned to Earth after a mission lasting 29 days 13 hours and 20 minutes.

Upon his return, Grechko was awarded the title of Hero of the Soviet Union and also received the Order of Lenin.

His second spaceflight came as flight-engineer onboard Soyuz 26, launched on 10 December 1977. Grechko and Yuri Romanenko docked with the Salyut 6 space-station, with the prime objective of the mission being an examination of the station's docking mechanism, as it was felt that it might have been damaged during Soyuz 25's abortive docking attempts. Soyuz 26 docked at the station's rear docking port, and Grechko and Romanenko carried out an EVA to thoroughly examine the suspect forward docking port. They could find no damage.

Romanenko was supposed to stay in the airlock while Grechko performed the EVA. However, he wanted to see the view from outside and performed an unauthorized exit of the space-station. Unfortunately, he lost his grip and according to one report, floated off. To his horror, he realized that his safety line had not been attached. Grechko, in his own words, was able to grab the line 'near the end' and haul Romanenko back in! The two men then re-entered the Salyut, and settled down to their long programme of scientific experiments and observations before returning to Earth after a mission lasting 96 days and 10 hours precisely. During the flight, the doctors apparently had to tell Grechko off for working too hard.

In making the flight, Grechko became the first astronaut or cosmonaut to fly two space-station missions. During the EVA, he spent 20 minutes outside the Salyut, and was in 'the near vacuum conditions of space' for 1 hour and 28 minutes, which is understood to mean hatch-opening to hatch-closing.

He was subsequently awarded the title of Hero of the Soviet Union and also received the Order of Lenin, both for the second time. In addition, he was awarded the title of Hero of the Czechoslovakian Socialist Republic. He had experienced chest pains following his return to Earth after the Soyuz 17 flight, but encountered no such problems at the end of the Soyuz 26 mission.

Soyuz 26 was an unscheduled mission, which was required only when Soyuz 25 failed. There is some evidence to suggest that at the time Grechko was actually in training as back-up commander for the forthcoming joint Soviet-GDR spaceflight. He was moved into the Soyuz 26 crew as he was a specialist in the Salyut docking mechanism, and his place on the GDR mission was taken by Viktor Gorbatko.

Grechko's next assignment was as CapCom during the EVA of Ivanchenkov and

Kovalyonok from Salyut 6 in July 1978, and in August 1979, he served as CapCom during the EVA of Lyakhov and Ryumin. He was named in October 1983 as back-up to Nikolai Rukavishikov, (later replaced by Gennady Strekalov), the flight-engineer for the joint Soviet-Indian mission, Soyuz T-11, launched on 2 April 1984. For his part in the preparations for the mission, Grechko was awarded India's Kirti Chakra Order.

When Salyut 7 suffered an onboard failure which left it without power in February 1985, a repair mission became necessary. This caused a disruption in the flight-schedule, *(See section entitled 'Salyut 7 Crewing' for more details)* and once the Salyut had been repaired, Grechko was brought in to serve as flight-engineer onboard Soyuz T-14, launched on 17 September 1985. *It is also believed that he served as the back-up flight-engineer on Soyuz T-13.* Together with Vladimir Vasyutin and Alexander Volkov, he linked up with the orbiting Salyut 7 space-station and joined the crew of Dzhanibekov and Savinykh already onboard. The two crews worked together for a week, and then Grechko returned to Earth with Dzhanibekov, the Soyuz T-13 commander, thus carrying out the first partial replacement of a crew. Vasyutin and Volkov remained onboard the Salyut with Viktor Savinykh, thus restoring the original Soyuz T-13 crew.

Grechko's flight had lasted 8 days 21 hours and 13 minutes, and upon his return, he was awarded his third Order of Lenin. During his stay onboard Salyut 7, he conducted studies of atmospheric phenomena using the SKIF spectrometer, and it seems likely that he also checked the repairs carried out to the station by Dzhanibekov and Savinykh. At 54, he was the oldest cosmonaut ever to fly in space.

He is married, and his wife Nina Viktorovna is an engineer with a degree in Technology, who works as a research worker in a design bureau in Moscow. They have two sons, Alexei (Alyosha) and Mikhail (Misha).

Grechko is a man of many hobbies, including mountain-skiing, underwater spear-fishing, motor racing and philately. He heads a motorists club in Moscow, is Chairman of the Soviet Mountain Skiing Federation, and is a member of the Soviet National Philatelic Society (the All Union Philatelic Society), which he joined in 1970. He specializes, not suprisingly, in stamps featuring space research, and his collection includes, of course, several issues devoted to his own spaceflights. He also collects stamps on medicine, geography, radio-physics, astro-physics and astronomy, related to space-research. There is apparently a philatelic club in Star Town, to which Grechko and other cosmonauts belong. He is also said to do a great deal of social work, and is Deputy Chairman of the Soviet Peace Committee.

Grechko is considered to be the best astro-physicist in the cosmonaut team, and he was awarded the Tsiolkovsky Gold Medal of the Soviet Academy of Sciences in April 1978 for his work onboard Salyut 4 in 1975. He is credited with several discoveries concerning the precise structure of the Earth's atmosphere, and is a Doctor of Technical Sciences, having been awarded a doctorate in Physics and Mathematics in 1984. His dissertation was devoted to optical studies of the atmosphere and ionosphere and of astrophysical objects from Salyut space-stations.

He does however, have a less serious side. During his stay onboard Salyut 6, he reported sightings of UFO's outside the space-station. He saw the mystery objects whilst performing gymnastic exercises, and his UFO report was quoted in lectures all over the world. But in April 1982, he came clean, and admitted in Izvestia that it had been a practical joke. The 'objects' were actually flakes of metallic foil which had been shaken off the space-station during his exercises !

Georgi Grechko was the 34th Soviet cosmonaut, and he has spent a total of 134 days 20 hours and 33 minutes in space.

Major-General Alexei Alexandrovich Gubarev

Alexei Gubarev was born on 29 March 1931, in the village of Gvardeitsy, in the Borsky Region, Kuibyshev District on the Volga. He became interested in flying at an early age, and on leaving school in 1950, he joined the Soviet Army. He then graduated from a Naval Air Force School in 1952, and joined the Soviet Air Force. In December 1952, he served with the Pacific Fleet air units engaged in supporting the Chinese and North Koreans in the Korean War.

In 1957, he joined the CPSU, and enrolled at the Red Banner (now Gagarin) Air Force Academy for an engineering course. He graduated in 1961 as a military pilot, first-class, and then served in various air units. He was serving on the Black Sea as a squadron-leader, attached to naval units, when he was selected to join the cosmonaut team in 1963. He reported for duty on 11 January that year.

He was paired with Georgi Grechko at the end of 1971, *and it is thought that the two men may have served as the support crew for the Cosmos 557 Salyut attempt in May 1973 (see section entitled 'Salyut 2/Cosmos 557 Crewing' for more details.)* Gubarev's first official assignment was as the back-up to Vasily Lazarev, the commander of Soyuz 12, launched on 27 September 1973, and his first spaceflight came as commander of Soyuz 17, launched on 10 January 1975. Gubarev and Grechko docked with the Salyut 4 space-station and spent nearly 30 days onboard, carrying out Earth-resources work and astronomical observations. Gubarev had some trouble adapting to the state of weightlessness, but this was mainly because he and Grechko did not follow the recommended sleep periods. The two men returned to Earth after a flight lasting 29 days 13 hours and 20 minutes.

Upon his return, Gubarev was awarded the title of Hero of the Soviet Union and also received the Order of Lenin. He was paired with Vladimir Remek in May 1977, and in the following August, the two men began joint training for the Soviet-Czechoslovakian mission. Gubarev made a second spaceflight as commander of the joint mission, Soyuz 28, launched on 2 March 1978. He and Remek docked with the orbiting Salyut 6 space-station, and joined the crew of Romanenko and Grechko already onboard. They then spent nearly seven days carrying out a series of joint Soviet-Czech experiments and observations, before returning to Earth after a flight lasting 7 days, 22 hours and 16 minutes.

Gubarev was subsequently awarded the title of Hero of the Soviet Union and also received the Order of Lenin, both for the second time. He was also awarded the title of Hero of the Czechoslovakian Socialist Republic. In April 1978, he was awarded the Tsiolkovsky Gold Medal of the Soviet Academy of Sciences for his work onboard Salyut 4 in 1975.

Alexei Gubarev is married, and his wife Nadezhda Alexeyevna is a metallurgical engineer in the technical library of Star Town. Their son Vladimir is a graduate of the Institute of Economics and Statistics in Moscow, and their daughter, Olga, is a graduate of the Lomonosov University in Moscow. During his Air Force career and cosmonaut training, Gubarev has logged over 1500 hours flying time in jets.

Yuri Gagarin once said of him 'He flies courageously and confidently, and is diligent and energetic.' He likes precision, and admits to having a 'rather explosive' temper. Together with Vladimir Remek, he has co-written a book entitled 'Porodneniye Orbitoy' ('International Orbit'), published in 1983.

Alexei Gubarev was the 33rd Soviet cosmonaut, and he has spent a total of 37 days 11 hours and 36 minutes in space.

Major-General Jugderdemidiin Gurragcha

Jugderdemidiin Gurragcha was born on 5 December 1947 in Rashant, in the Gurvan-Bulak settlement of the Bulgan Province, Mongolia. His father was a peasant cattle breeder, who had never been to school, whereas his mother had only four years schooling. True to Mongolian tradition, Gurragcha saddled his first horse whilst still a child.

He graduated from secondary school in Bulgan and then enrolled at a college of agriculture in Ulan Bator, the capital of Mongolia. In 1968, he was conscripted, and served as a wireless operator in the Mongolian Army. In 1971, he was sent to study at a school for junior aviation specialists in the Soviet Union, and he completed the course with a qualification as a helicopter radio communications equipment mechanic. In 1973, he enrolled at the Zhukovsky Military Air Academy in the Soviet Union, and graduated in 1977. He then worked as an engineer on aircraft equipment at an air squadron.

Following the agreement to fly non-Soviet cosmonauts from member countries of the Intercosmos programme onboard Soyuz and Salyut spacecraft, Mongolia began a cosmonaut selection programme. The two finalists were Gurragcha and Majdarzavyn Ganzorig, who began their training at the Yuri Gagarin Cosmonaut Training Centre in April 1978.

Gurragcha was promoted to Captain in 1978, and in 1979, he joined the Mongolian People's Revolutionary Party. After initial training, he was teamed up with Vladimir Dzhanibekov, and eventually flew as research-cosmonaut onboard Soyuz 39, launched on 22 March 1981. He and Dzhanibekov linked up with the Salyut 6 space-station and joined the crew of Kovalyonok and Savinykh already onboard. The two men spent nearly seven days carrying out a programme of joint Soviet-Mongolian experiments, before returning to Earth after a flight lasting 7 days 20 hours and 43 minutes.

Upon his return, Gurragcha was awarded the title of Hero of the Soviet Union and also received the Order of Lenin. In addition, he was awarded the title of Hero of the People's Republic of Mongolia, and also received the Order of Suhbaatar and Gold Star Medal.

Known to his friends as Gurr, he is married, and he and his wife Batmunkh, a secondary school teacher, have a son, born in 1977 or 1978. Dzhanibekov once said of Gurragcha 'He's modest, reserved and has a sense of humour . . . a very skilled engineer.'

Jugderdemidiin Gurragcha was the 8th Intercosmos cosmonaut.

Munir Habib

Munir Habib is one of the two Syrian cosmonauts chosen for the joint Soviet-Syrian spaceflight.

He is a pilot in the Syrian Air Force, and began training at the Yuri Gagarin Cosmonaut Training Centre in late 1985.

At the time of his selection he was aged 32. He is married, and he and his wife Yumna have two sons, Mdyan, born in 1976 or 1977, and Raid, born in 1980 or 1981. No other biographical details have been released.

Colonel Miroslaw Hermaszewski

Miroslaw Hermaszewski was born on 15 September 1941, in Lipniki, in S.W. Poland, one of 7 children, with 4 sisters and 2 brothers. His father was a peasant small-holder, and was killed during World War Two, following which the family moved to Wolowo near Wroclaw. Aircraft modelling became a hobby of Miroslaw at primary school, and he became interested in aviation when his elder brother enrolled at the Deblin Air School in 1948. He was granted a glider-pilot's licence at the age of 17.

After graduation from school, he entered the Officers Flying School at Deblin in 1961. He graduated in November 1964, taking first place in the final examinations. He then served as a fighter pilot in the Polish Air Defence Force, and has logged almost 1500 hours in the air.

From 1969 to 1971, he attended a three year course in the aviation department of the Karol Swierczewski General Staff Academy, at the Faculty of Science. He graduated in 1971, and was promoted to Captain, and then held a number of commanding posts in various Air Defence Force Units. In January 1975, he was promoted to Major, and made a squadron leader, and in April 1976, he was appointed commander of a fighter regiment in the Polish Air Force. For his flawless service he was awarded the Order of the Gold Crest of Merit, together with a series of medals.

While at school, and during his military service, he was an activist of the Socialist Youth Union, and in 1963, he joined the PZPR – the Polish United Workers Party.

Following the agreement to fly non-Soviet cosmonauts from member countries of the Intercosmos programme onboard Soyuz and Salyut spacecraft, Poland began a cosmonaut selection programme. The two finalists were Hermaszewski and Zenon Jankowski, who began their training at the Yuri Gagarin Cosmonaut Training Centre in December 1976.

In August 1977, he began training with Pyotr Klimuk, but this training was interrupted when Hermaszewski was hospitalized with tonsilitis, and had to undergo a minor operation. He made a speedy recovery, and quickly resumed training.

He eventually flew as research-cosmonaut onboard Soyuz 30, launched on 27 June 1978.

He and Klimuk linked up with the Salyut 6 space-station and joined the crew of Kovalyonok and Ivanchenkov already onboard. The two men carried out a programme of joint Soviet-Polish experiments and observations before returning to Earth after a mission lasting 7 days 22 hours and 03 minutes.

Upon his return, Hermaszewski was promoted to Lieutenant-Colonel, and was awarded the Grunwald Cross, 1st Class, by the Polish Council of State, and the Polish Academy of Sciences's N. Kopernikus medal. In addition, he was awarded the title of Hero of the Soviet Union and also received the Order of Lenin.

According to unofficial philatelic information from Poland, Zenon Jankowski, the back-up on Soyuz 30, was in fact, chosen by the Soviet Union to be the first Pole in space. The Polish Government therefore printed two stamps honouring cosmonaut Jankowski as part of the 1978 Intercosmos programme.

However, the day before the flight, the Polish Philatelic Agency, Ars Polona, received word that Miroslaw Hermaszewski was to be the first Polish cosmonaut. New stamps were immediately printed, and the incorrect stamps were destroyed, but some apparently escaped the shredding machines.

When martial law was imposed in Poland in December 1981, a Military Committee for National Salvation was formed, and Hermaszewski was one of the 20 military officers chosen to serve on it. In 1982, he graduated from the Voroshilov High Command Academy.

He married in 1965, and his wife, Emilia, born on 23 August, now works for Flot, the Polish airline. They have a son, Miroslaw, born in 1966 or 1967, and a daughter, Emilia, born in 1974 or 1975. Hermaszewski's hobbies include reading SF books, gliding, track and field athletics, and hunting. He is also a nature lover.

Miroslaw Hermaszewski was the 2nd Intercosmos cosmonaut.

Valery Vasilyevich Illarionov

NO PHOTOGRAPH AVAILABLE

During the preparations for the Apollo-Soyuz Test Project, the Soviet Union appointed four CapComs. They were Georgi Shonin, Vladimir Dzhanibekov, Yuri Romanenko – all cosmonauts – and a fourth man, Valery Illarionov. The last-named had been assigned as the CapCom to be present in the Soviet support room at the Johnson Space Centre, Houston, during the ASTP mission. The NASA Public Affairs Office at JSC confirmed, in private correspondence with the author at the time, that Illarionov was indeed a cosmonaut himself.

The Soviet Academy of Sciences later officially confirmed the identity of Illarionov, but declined to give further personal details, saying only that he was born on 2 July 1939, and was a civilian cosmonaut selected in 1970.

It is believed that following the completion of the ASTP mission, Illarionov may have trained for a Salyut flight in the late '70's, as did the other cosmonauts selected in 1970. He may have been given a Salyut 6 assignment, but it would appear that at some stage, he became physically disqualified and was replaced.

All the 1970 cosmonauts assigned to ASTP have now made a spaceflight, with the exception of Boris Andreyev and Illarionov. He was still at Star Town in November 1984, but it is not known if he is still active.

There is also some doubt that he is a civilian engineer, and it is now believed that he is a military pilot.

Alexander Sergeievich Ivanchenkov

Alexander Ivanchenkov was born on 28 September 1940, in Ivanteyevka, a small town near Moscow. His father, a civil engineer, was killed in 1942, during World War 2, and his mother died in 1947, and he was therefore brought up by his grandmother and an aunt. Upon finishing secondary school with a Gold Medal in 1958, he went to the Moscow Aviation Institute. While there, he designed systems for sports planes and gliders, and became a member of the student's scientific society. He graduated in 1964, and went to work in the Korolyov design bureau, where he participated in the development of new space vehicles and rapidly moved ahead in his work on engineering computations and computer technology.

He was selected to join the cosmonaut team in 1970, when he was said to be a well known aircraft designer and pilot. He joined the CPSU in 1972, and his first assignment came in May 1973, when he was named as back-up to Nikolai Rukavishnikov, the flight-engineer of the reserve Soyuz for the Soviet half of the Apollo-Soyuz Test Project. From May 1973 he trained for the mission in both the United States and the Soviet Union. As part of the preparations for the joint flight, he served as back-up to Rukavishnikov on the Soyuz 16 ASTP rehearsal mission, launched on 2 December 1974.

Although the identity of the back-up flight-engineer for Soyuz 25 has never been revealed, it is widely believed that it was Ivanchenkov. He should then have flown as flight-engineer on the next mission, but instead had to serve as back-up again. (see section entitled 'Salyut 6 Crewing' for more details.)

His next assignment therefore was as the back-up to Georgi Grechko, the flight-engineer onboard Soyuz 26, launched on 10 December 1977, and then as the back-up to Oleg Makarov, the flight-engineer onboard Soyuz 27, launched on 10 January 1978. His first spaceflight came as flight-engineer onboard Soyuz 29, launched on 15 June 1978. He and Vladimir Kovalyonok linked up with the Salyut 6 space-station, and carried out a long programme of research and experiments as well as playing host to the visiting cosmonauts from Poland and the German Democratic Republic.

During their stay onboard the Salyut, Kovalyonok and Ivanchenkov made an EVA lasting 2 hours and 5 minutes. Kovalyonok stayed in the hatchway while Ivanchenkov removed samples from the Medusa experiment and replaced them with fresh materials. The two men also tested the new spacesuits.

They returned to Earth after a flight lasting 139 days 14 hours and 48 minutes – thus setting a new duration record.

Upon his return, Ivanchenkov was awarded the title of Hero of the Soviet Union and also received the Order of Lenin. In addition, he was awarded the title of Hero of the German Democratic Republic with the Order of Karl Marx, and received the Polish Gruenwald Cross, 1st Class.

His next assignment was as the back-up to Oleg Makarov, the flight-engineer onboard Soyuz T-3, launched on 27 November 1980, and in October 1981, he was named as the prime flight-engineer for the joint Soviet-French spaceflight, Soyuz T-6, launched on 24 June 1982.

Vladimir Dzhanibekov, Jean-Loup Chrétien and Ivanchenkov linked up with Salyut 7, and joined the crew of Berezovoi and Lebedev already onboard. They spent nearly seven days carrying out a programme of joint Soviet-French experiments, before returning to Earth after a flight lasting 7 days 21 hours and 51 minutes. Upon his return, Ivanchenkov was awarded the title of Hero of the Soviet Union and also received the Order of Lenin, both for the second time. He also received the French Order of the Legion of Honour.

Ivanchenkov, known as Sasha to his friends, is married, and his wife Rimma Alexandrovna is a medical internist, having completed a graduate course at the First Medical Institute. They have a daughter named Tanya. Ivanchenkov has been described as quite introverted, yet quite phlegmatic. His hobbies include travel, skiing, basketball, tennis and playing the guitar. He has played at a pop concert, and during his long stay onboard Salyut 6, was pleasantly suprised when a Progress ferry delivered a guitar to him as a recreational gift from FCC. During the mission, he had also celebrated his 38th birthday.

Alexander Ivanchenkov was the 44th Soviet cosmonaut, and he has spent a total of 147 days 12 hours and 39 minutes in space.

Colonel-Engineer Georgi Ivanov Ivanov

Georgi Ivanov was born on 2 July 1940, in the town of Lovech, in Bulgaria. His father Ivan Ivanov is a retired electrical technician with the Power Supply Administration of Bulgaria, while his mother Anastasia Georgieva is a housewife. In 1955, Ivanov became interested in flying, and within a year he was among the most active members of the local model flying club, and had begun making parachute jumps. In 1957, he completed the initial course in gliding at the airclub in Preven, and in less than a year was piloting a plane.

He finished school in 1958 and enrolled in the Georgi Benkovski Higher Air Force School near Dolna Mitropoliya, not far from Preven. He graduated in 1964 as a Lieutenant, with the diploma of engineer-pilot, and became an instructor at the School. In 1967, he was promoted to Coach-pilot 2nd class, and in 1968, he became a Coach-pilot 1st class. He also joined the Bulgarian Communist Party in 1968. Ivanov later served in the anti-aircraft units of the Bulgarian Air Force as a flight commander and squadron leader, and was awarded the title of Military pilot 1st class, together with medals for exemplary service. He has logged more than 1900 hours flying time.

Following the agreement to fly non-Soviet cosmonauts from member countries of the Intercosmos programme onboard Soyuz and Salyut spacecraft, Bulgaria began a cosmonaut selection programme. The two finalists were Ivanov and Alexander Alexandrov, and they were sent to begin their training at the Yuri Gagarin Cosmonaut Training Centre in March 1978. Ivanov eventually flew the mission as cosmonaut-researcher onboard Soyuz 33, launched on 10 April 1979. He and Nikolai Rukavishnikov were scheduled to dock with the Salyut 6 space-station and spend a week visiting the crew already onboard. However, the two men were forced to return to Earth prematurely when the Soyuz approach engine malfunctioned during the final stages of the rendezvous manoeuvres. Following an emergency ballistic re-entry, they returned to Earth after a flight lasting only 1 day 23 hours and 1 minute.

It was not until October 1983 that the true extent of the failure and the peril that the crew had been in, was revealed. The failure occurred in the main 917lb thrust rocket engine used for rendezvous manoeuvres and the re-entry burn. After it had been fired six times during rendezvous manouevres, its seventh and last planned firing shut down after only 3 seconds of a planned 6 second burn.

Ground control studied the problem, and decided that the engine should be fired again in an attempt to complete the rendezvous with Salyut 6. However, when it was re-started it shut down immediately, and the crew reported feeling unusual vibrations during the firing sequence. At the time, the Soyuz was only 3000 feet (914 m) from the space-station, and the Salyut cosmonauts reported that they had seen the engine fire, but with 'a lateral glow in the power compartment.'

This report caused great alarm at ground control, as it was feared that the abnormal plume could have caused serious internal damage to the 904lb thrust back-up engine located next to the main engine. If the back-up engine failed to work, it would mean that no re-entry burn was possible, and the crew would be marooned in orbit. Calculations showed that Soyuz 33's orbit would decay naturally after 10 days, but the crew had less than 5 days of oxygen remaining.

After a worrying night considering all the possibilities, the decision was taken to attempt a re-entry burn using the back-up engine. During the 188 second burn, the engine malfunctioned as expected, but instead of shutting down early, it kept burning beyond the planned cut-off point, and started firing at less than full thrust. Eventually, after 25 seconds of excessive burn time, the crew shut down the engine manually.

The overburn was serious, as it could have resulted in a dangerously steep re-entry angle, during which the descent module might burn up. In the event, Soyuz 33 flew a highly stressed ballistic re-entry in which the cosmonauts experienced 8-10g loads compared with the 3-4g's experienced on normal re-entries. As a result of the overburn, the descent module missed the landing window, falling short of the touchdown site by several hundred miles.

Even though the mission was a failure, Ivanov returned to a hero's welcome, and received many awards and honours, including:- the title of Hero of the Soviet Union, with the Order of Lenin; the title of Hero of the People's Republic of Bulgaria, with Gold Star Medal and the Order of Georgi Dimitrov; the title of pilot-cosmonaut of the People's Republic of Bulgaria; the title of Merited-pilot of the People's Republic of Bulgaria; the Honorary Badge of Sofia, 1st Class, and the Honorary Badge of Botevgrad, Bulgaria. In addition, he was made an honorary citizen of Omourtag and Lovech, Bulgaria. He was also promoted from Major to Lieutenant-Colonel.

Known to his friends as Gosho (an abbreviation of Georgi), Ivanov is married, and he and his wife – Natalia Rousanova, who works in the public service sector – have a daughter, Annie (or Ani) born in 1966 or 1967.

The psychiatric reports carried out by Bulgarian specialists during preparations for the mission said of Ivanov 'Definitely positive traits, stable and mature character. Strong, mobile, even-tempered type of nervous system. High emotional, moral and psychological stability, excellent flying qualities, rich flying experience. Perfect health and fitness, high physical endurance.'

Rukavishnikov once said of him 'When I first met him, he seemed to me to be somewhat morose. It won't be easy to work with him, I thought. But this first impression proved wrong. Georgi is a jolly good fellow, a cordial friend. He appeared to me to be very serious, gloomy, not once did he smile – a really serious man, but also with a sense of humour. Careful, reliable, a tireless worker. Strong-willed and determined.'

A former colleague, Vesyelin Stoyanov, said 'I studied together with him in flying school and subsequently served in the Army. On jets, he flew about 1900 hours and was always one of the first to master new techniques. One of his main qualities is to thoroughly study everything and not disregard small things.'

His hobbies include reading and underwater fishing, and he is also a good skier. Following his spaceflight, he was apparently forbidden to fly by the Minister of National Defence, whose own son, Captain-Engineer Chavdar Djurov – another Bulgarian candidate cosmonaut – was killed in a jet-fighter crash. However, Ivanov was eventually allowed to fly again in MiG.-21's.

Georgi Ivanov was the 4th Intercosmos cosmonaut.

Colonel Sigmund Jähn

Sigmund Jähn was born on 13 February 1937, in Rautenkranz, near Klingenthal in the Vogtland of the German Democratic Republic, the son of a sawmill worker. His mother contributed to the upkeep of the family by taking in sewing and taking up occasional employment with the forestry authority. At school, he was elected Pioneer class organizer, and became a member of the Free German Youth (FDJ), of which he later became Secretary. After finishing 8th Grade with the top marks of his year, he left school in 1951 to learn the trade of printer, and he received his skilled workers certificate from the Falkenstein national book printing works, Klingenthal branch.

In 1955, Jähn joined the Air Force of the National Peoples Army. He attended the Franz Mehring Air Force Defence Officers College, and was described as a model officer cadet. In 1956, he joined the Socialist Unity Party, and in 1958, he was commissioned as a 2nd Lieutenant.

He was assigned to a fighter squadron where he was the political officer responsible for the ideological education of the members of his unit. At the age of 26, he was appointed officer in charge of air tactics and gunnery in his squadron.

In 1966, he was sent for training at the Gagarin Air Force Academy in the Soviet Union, and out of 21 training subjects, he received 13 'Excellent' ratings, and 8 'Good.' He graduated in 1970, and returned to the GDR. In the period that followed, he assumed a number of important posts in the Air Force, and in no time at all, had again attained top class flying qualifications.

He received his licence as a flight instructor for day and night training under all weather conditions, and won particular distinction for his theoretical development and practical testing of new elements in the training and use of fighter forces. For his outstanding services, he was awarded the title of Distinguished Military Pilot of the GDR. He has logged over 1000 hours of flying time.

Following the agreement to fly non-Soviet cosmonauts from member countries of the Intercosmos programme onboard Soyuz and Salyut spacecraft, the German Democratic Republic began a cosmonaut selection programme. The two finalists were Jähn and Eberhard Köllner, and they were sent to begin their training at the Yuri Gagarin Cosmonaut Training Centre in December 1976.

Jähn was paired with Valery Bykovsky in May 1977, and eventually flew as research-cosmonaut onboard Soyuz 31, launched on 26 August 1978. He and Bykovsky docked with the orbiting Salyut 6 space-station, and joined the crew of Kovalyonok and Ivanchenkov already onboard. The two men then carried out a joint programme of Soviet-GDR experiments and observations before returning to Earth after a flight lasting 7 days 20 hours and 49 minutes.

Upon his return, Jähn was awarded the title of Hero of the Soviet Union and also received the Order of Lenin. He was also awarded the title of Hero of the German Democratic Republic, and was promoted from Lieutenant-Colonel to Colonel. In addition, he received the following honours and awards:- the title of Pilot-Cosmonaut of the GDR, the Order of Karl Marx, the F.Gekkert Gold Medal (GDR), the GDR Academy of Sciences Leibnitz Medal, the Arthur Becker Gold Medal of the FDJ – the Free German Youth, and the freedom of Berlin.

Jähn is married, and his wife Ericka trained first as a fitter and then qualified as a technical draughtsman. They have two daughters, Marina, born in 1958, – a typist – and Grit, born in 1966. Jähn's favourite hobby is reading, and he is also a passionate hunter, being a member of a German hunting club. He is also a nature lover.

He has been described as 'strong-willed, single-minded, resolute, and has the capacity to react swiftly in unforeseen circumstances.' Alexei Leonov once described him as 'an unusually uncomplicated man – a man of great principles and integrity.'

Sigmund Jähn was the 3rd Intercosmos cosmonaut.

Colonel Zenon Jankowski

Zenon Jankowski was born on 22 November 1937, in Poznan, Poland. Following graduation from school, he entered an Air Force Officers School in 1956. He has also studied at the Karol Swierczewski General Staff Academy, and joined the PZPR – the Polish United Workers Party – in 1965. In 1976, he commanded a group of variable-geometry aircraft.

Following the agreement to fly non-Soviet cosmonauts from member countries of the Intercosmos programme onboard Soyuz and Salyut spacecraft, Poland began a cosmonaut selection programme. Jankowski's 'great technical knowledge, flying experience, tough mental build-up, and physical strength' led him into the final group of cosmonaut candidates. The two finalists were Jankowski and Miroslaw Hermaszewski, who began their training at the Yuri Gagarin Cosmonaut Training Centre in December 1976.

Hermaszewski eventually flew as research-cosmonaut onboard Soyuz 30, launched on 27 June 1978, and Jankowski served as back-up. Following a successful launch, he worked in FCC acting as a 'consultant to the Senior Ground Flight Controller' i.e. probably CapCom.

Following the completion of the flight, Jankowski was promoted from Lieutenant-Colonel to Colonel, and was awarded the Commanders Cross of the Polonia Restituta (otherwise known as the Commodore Cross of the Order of Poland's Resurgence.) This was in recognition of 'his services in the preparation of the flight and exemplary fulfilment of his tasks as consultant to the Flight Controller.'

According to unofficial philatelic information from Poland, Jankowski was in fact chosen by the Soviet Union to be the first Pole in space. The Polish Government therefore printed two stamps honouring him as part of the 1978 Intercosmos programme.

However, the day before the flight, the Polish Philatelic Agency, Ars Polona, received word that Miroslaw Hermaszewski was to be the first Polish cosmonaut. New stamps were immediately printed, and the incorrect stamps were destroyed, but some apparently escaped the shredding machines.

Wing Commander Jankowski is married, and has a daughter, born in 1964 or 1965.

Colonel Yevgeny Vasilyevich Khrunov

Yevgeny Khrunov was born on 10 September 1933, in Prudy, a village in the Tula Region, into a peasant farmers family. His father Vasily Khrunov died in 1948, and his mother Agrafena Nikolayevna was left with eight children to bring up. Yevgeny left a 7 year school and went on to an agricultural technical training school in Kashira. He could have become an expert on farm machinery, but for a sudden passion for flying. In 1952, he joined a pilots school, going onto the Bataisk Military Aviation College in 1953, and graduating in 1956.

From 1956 onwards, he served as a pilot in the Soviet Air Force, and apparently stood out for his flying abilities, his love of sports, and his 'remarkable ability' with machines. At one time, he served in the same air regiment as Viktor Gorbatko. He joined the CPSU in 1959, and he was selected to join the cosmonaut team in 1960.

His first assignment was as back-up to Alexei Leonov, the co-pilot onboard Voskhod 2, launched on 18 March 1965, and he was therefore EVA-trained. While continuing his training, Khrunov studied at the Zhukovsky Military Air Academy, and graduated in 1968 with honours.

On 23 April 1967, Vladimir Komarov was launched onboard Soyuz 1. He was brought back to Earth after just over a day in orbit, and was killed when the spacecraft's parachute tangled following re-entry. It is now clear, despite official Soviet denials at the time, that a second spacecraft, Soyuz 2, was to have been launched on 24 April. Following docking, an exchange of crews would have taken place.

The Soyuz 2 prime and back-up crews have since been identified by Western observers. The prime crew were Valery Bykovsky, Alexei Yeliseyev and Yevgeny Khrunov, and the back-up crew were Pavel Belyayev/Andrian Nikolayev, Valery Kubasov and Viktor Gorbatko. Khrunov's next assignment therefore was as test-engineer for Soyuz 2, and he would have made an EVA to transfer from Soyuz 2 to Soyuz 1. However, following the initial problems with Soyuz 1, the Soyuz 2 launch was cancelled.

An unmanned Soyuz 2 subsequently flew in a rendezvous operation with Soyuz 3, piloted by Georgi Beregovoi, and in January 1969, nearly two years late, the mission intended for Soyuz 1/2 was carried out by the Soyuz 4/5 spacecraft. The prime and back-up crews of Soyuz 2 were given the same assignments on Soyuz 5 – only the commanders had changed.

Khrunov's first spaceflight therefore came as test-engineer onboard Soyuz 5, launched on 15 January 1969. Soyuz 5 docked with Soyuz 4, carrying Vladimir Shatalov, in the world's first docking of two manned spacecraft. Khrunov and Alexei Yeliseyev then carried out an EVA to transfer from Soyuz 5 to Soyuz 4, and joined Shatalov inside. The EVA lasted 1 hour, of which 15 minutes was spent in open space. The two craft then separated, and the next day Soyuz 4 re-entered, Khrunov having spent 1 day 23 hours and 39 minutes in space.

Upon his return, Khrunov was awarded the title of Hero of the Soviet Union and also received the Order of Lenin. He began studying, and went on to earn a Candidate of Technical Sciences degree, and in 1974, he graduated from the V.I. Lenin Military-Political Academy. Also in 1974, he served as CapCom on the Soyuz 14/Salyut 3 mission. He also served as a CapCom on Salyut 5, and in 1978, he served as CapCom on the Soyuz 30 mission.

In September 1979, he teamed up with the Cuban, José Lopez-Falcon in preparation for the forthcoming joint Soviet-Cuban mission, and he served as back-up to Yuri Romanenko who flew as commander of the joint flight, Soyuz 38, launched on 18 September 1980. *According to several reports, when Valery Bykovsky became disqualified, Khrunov was offered the position of commander for the joint Soviet-Romanian spaceflight, but he declined. In the event, the mission, Soyuz 40, launched on 14 May 1981, was commanded by Leonid Popov. (see section entitled 'Intercosmos Crewing' for more details.)*

Khrunov has co-authored two books, 'Man as Operator in Cosmic Flight', and a science-fiction novel 'The Path To Mars' published in 1979. He is married, and he and his wife Svetlana Khrunova, a secondary school teacher, have a son, Valery. Known to his friends as Zhenka, an abbreviation of Yevgeny, Khrunov likes all sports, but is particularly keen on gymnastics and tennis. He is a regular theatre-goer, enjoys reading, and is Chairman of the Soviet Federation of Helicopter Sport. He has an older brother, Vladimir.

Yevgeny Khrunov was the 16th Soviet cosmonaut.

Colonel Leonid Denisovich Kizim

Leonid Kizim was born on 5 August 1941, in the city of Krasny Liman in the Donetsk Region of the Ukraine, where his parents both worked on the railway. Kizim was apparently 7 years old when he first said that he wanted to become a pilot. After finishing secondary school in 1958, he applied for enrolment in a flying school, but was turned down because he lacked the required height and weight. He refused to give up, and tried again in 1959. This time he was successful, and was admitted to the Higher Air Force College, in Chernigov, in the Ukraine. He graduated with honours in 1963, and then joined the Soviet Air Force. He served in units in the Caucasus, and during his service career, he tested 12 different types of aircraft.

It is said that Kizim's Air Force colleagues noted his 'ability to handle sophisticated mechanisms, his perseverance and discipline', and after two years of service it was suggested that he apply to join the cosmonaut team. He was passed by a selection committee in October/November 1965, and joined the CPSU in 1966.

Following his selection, he learned to fly several types of aircraft and was awarded the qualification of Test-Pilot 3rd Class. He is also a military pilot 1st Class, and has logged more than 1600 hours flying time.

It appears that Kizim may have been involved in some way with the Soyuz 6, 7 and 8 missions, all launched in October 1969, possibly as a CapCom.

He graduated from the Gagarin Air Force Academy in 1975, through a correspondence course, and in the same year, he was assigned to a Salyut training group which also included Popov and Romanenko.

Kizim's first assignment was as the back-up to Yuri Malyshev, the commander of Soyuz T-2, launched on 5 June 1980, and his first spaceflight came as commander of Soyuz T-3, launched on 27 November 1980. The mission was the first test-flight of the three man version of the Soyuz T. The spacecraft docked with the Salyut 6 space-station, and Kizim, Gennady Strekalov and Oleg Makarov spent 10 days carrying out repairs to the station before returning to Earth after a mission lasting 12 days 19 hours and 8 minutes.

Upon his return, Kizim was awarded the title of Hero of the Soviet Union and also received the Order of Lenin. According to Soviet reports, he had been actively involved in the designing and ground testing of the Soyuz-T spacecraft.

His next assignment came in October 1981, when he was named as back-up to Yuri Malyshev (later replaced by Vladimir Dzhanibekov), the commander of the joint Soviet-French mission, Soyuz T-6, launched on 24 June 1982. For his part in the preparations for the mission, he was awarded the French Order of the Legion of Honour.

Following the Soyuz T-6 mission, Kizim joined the special Salyut training group preparing for long-duration missions.

It is thought that his next assignment was as the commander of the support crew for Soyuz T-8, launched on 20 April 1983. It is believed that he and Vladimir Solovyov were then given the back-up assignment for Soyuz T-9, launched on 27 June 1983. During Lyakhov and Alexandrov's 2 EVA's from Salyut 7 in November 1983, Kizim and Solovyov conducted similar activities in the water tank at the Yuri Gagarin Cosmonaut Training Centre.

The two men should have been the prime crew on the next mission, but because of the 're-cycling' of Titov and Strekalov, they again served as the back-up crew. (See section entitled 'Salyut 7 Crewing' for more details.)

Kizim's next assignment therefore came as back-up to Vladimir Titov, the commander of Soyuz T-10, launched on 27 September 1983. The mission was aborted following a launch pad fire, and Kizim and Vladimir Solovyov were re-assigned as prime crew for the replacement flight. The aborted mission was never given a designation, and Kizim, Solovyov and Oleg Atkov were launched onboard the new Soyuz T-10 on 8 February 1984.

The three men docked their spacecraft with the orbiting Salyut 7 space-station and carried out a long programme of research and experiments as well as playing host to the joint Soviet-Indian spaceflight Soyuz T-11, and the visiting Soyuz T-12.

A major task carried out by Kizim and Solovyov was a series of six EVA's during which they completed extensive and complicated repairs to the Salyut.

The first four EVA's were concerned with the fuel system. The two men opened the side of Salyut 7 and replaced an oxidizer line which had leaked, damaging the Salyut propulsion system. They also installed a new but separate parallel propellant line in the area where the failure occurred.

The four EVA's took place in the period 23 April – 4 May, and lasted 4 hours and 15 minutes; 4 hours and 56 minutes; 2 hours and 45 minutes; and 2 hours and 45 minutes respectively.

A fifth EVA lasting 3 hours and 5 minutes was conducted on 18 May, during which two extra solar panels were installed on one of Salyut 7's three main arrays.

One more EVA was scheduled to complete the repairs to the fuel system. Vladimir Dzhanibekov had undergone special training in the water tank at the Yuri Gagarin Cosmonaut Training Centre and had practised with the tools and procedures necessary to compress and seal the ruptured fuel line. It was originally intended that he would carry out the repair himself during the Soyuz T-12 visit. However, Kizim and Solovyov, having carried out the four preparatory EVA's, asked to complete the job themselves. Therefore,

Dzhanibekov delivered a section of pipeline and a video film of his water tank repair procedure, and showed how the repairs should be carried out.

Kizim and Solovyov conducted the final EVA lasting 5 hours on 8 August, during which they compressed and sealed of the fuel line with the aid of a special tool. They had spent a total of 22 hours and 46 minutes conducting the six EVA's.

The three men resumed their normal programme of research for a further two months, before returning to Earth after a flight lasting 236 days 22 hours and 50 minutes – thus setting a new duration record.

Georgi Beregovoi later revealed how Kizim felt upon his return: 'Poor Leonid! He now understands the hardship of our civilization! Basically, he felt as though he had been beaten by a huge, large but soft bag. He was in pain all over – pressed down by some huge weight.'

Kizim was subsequently awarded the title of Hero of the Soviet Union and also received the Order of Lenin, both for the second time. For his part in the joint Soviet-Indian mission, he was also awarded the Indian Order of Kirti Chakra.

Known to his friends as Lenya, Kizim is married, and his wife Galina Fedorovna is an engineer working at a design office in Zvezdny Gorodok. He was already in the cosmonaut team when they married, and they have a son, Leonid, born in 1973, and a daughter, Tatyana, born on 24 May 1984, whilst Kizim was onboard Salyut 7. The first he saw of her was in photographs and video tapes brought up to the space-station by the visiting Soyuz T-12 crew, when she was two months old.

His hobbies include touring by car, and he has been described by Baudry as 'a lively and cheerful man, always in good shape.'

Leonid Kizim was the 48th Soviet cosmonaut, and he has spent a total of 249 days 17 hours and 58 minutes in space.

Major-General Pyotr Ilyich Klimuk

Pyotr Klimuk was born on 10 July 1942, in the village of Komarovka, near Brest in Byelorussia, where his mother Marfa still lives. He was born during the Nazi occupation of the area, and his father, Ilya Fyodorovitch Klimuk , a farmer, was killed during fighting for the liberation of Poland in the Battle for Magnuszew Beachhead in December 1944.

Klimuk was apparently something of a child genius, and after finishing school, he went to the Higher Air Force College in Chernigov, in the Ukraine. He graduated in 1964, and joined the Air Force. In 1963, at the age of only 20, he was accepted into the CPSU, which was a nearly unprecedented honour.

In October 1965, he was selected to join the cosmonaut team, at the age of only 23. He completed basic training after two years, and in mid-1967, he was assigned to the Zond training group, and began preparing for a flight to the Moon. Upon the abandonment of the Soviet lunar programme, he was re-assigned to the Soyuz programme, and his first assignment was as commander of Soyuz 13, launched on 18 December 1973. During the flight, Klimuk and flight-engineer Valentin Lebedev carried out studies of the stars using the Orion 2 telescope array. They also carried out medical experiments and tested new spacecraft systems before returning to Earth after 7 days 20 hours and 56 minutes in space.

Upon his return, Klimuk was awarded the title of Hero of the Soviet Union and also received the Order of Lenin. Prior to his spaceflight, he was studying at the corresponding department of the Gagarin Air Force Academy, and he returned to this following the mission.

Klimuk had been teamed with Vitaly Sevastyanov in early 1972, and later flew a long-duration mission with him in 1975. It is puzzling therefore that he flew with Lebedev on Soyuz 13, particularly as Lebedev had only been a cosmonaut for a year, and had not performed any back-up duties. In addition, Lebedev is known to have begun his training with Vladimir Kovalyonok. It is therefore possible to speculate that Sevastyanov was originally assigned to fly with Klimuk and that he was replaced by his back-up – Lebedev – at the last-minute when he became disqualified for some reason. It is interesting to note that the Soyuz 13 back-up crew has never been identified.

It is believed that Klimuk's next assignment was as the back-up to Alexei Gubarev, the commander of Soyuz 17, launched on 10 January 1975. (see section entitled 'Salyut 4 Crewing' for more details.)

He then served as the back-up to Vasily Lazarev, the commander of Soyuz 18, launched on 5 April 1975. Following a booster malfunction during launch, the mission was aborted, and was subsequently designated the 'April 5 anomaly.' Klimuk then flew on the replacement mission, termed Soyuz 18, launched on 24 May 1975.

The spacecraft linked up with the Salyut 4 space-station, and Klimuk and Vitaly Sevastyanov spent two months onboard conducting a wide range of technological and biological experiments, together with Earth observations. They then undocked and returned to Earth after a flight lasting 62 days, 23 hours and 20 minutes.

Klimuk was subsequently awarded the title of Hero of the Soviet Union and also received the Order of Lenin, both for the second time.

He graduated from the Commanding Department of the Gagarin Air Force Academy in 1977, and in August of that year, he began training with Miroslaw Hermaszewski, the Polish cosmonaut, for the forthcoming joint Soviet-Polish mission. In early 1978, he was amongst those in charge of the Soyuz 26 flight, and later that year, he was appointed Deputy Head-Political Affairs of the Yuri Gagarin Cosmonaut Training Centre.

His third spaceflight came as commander of Soyuz 30, launched on 27 June 1978. He and Miroslaw Hermaszewski docked with the Salyut 6 space-station and joined the crew of Kovalyonok and Ivanchenkov already onboard. The two men carried out a programme of joint Soviet-Polish experiments and observations before returning to Earth after a mission lasting 7 days 22 hours and 03 minutes.

Upon his return, Klimuk was awarded the Grunwald Cross, First Class by the Polish State Council, the N. Kopernikus medal by the Polish Academy of Sciences, and his third Order of Lenin. He had also been awarded the Tsiolkovsky Gold Medal in April 1978 for his work onboard Salyut 4 in 1975. Prior to the Soyuz 30 mission, it was stated that he held six medals in addition to his two Orders of Lenin.

In late 1982, it was reported that he and Alexei Leonov were both Deputy Directors of the Yuri Gagarin Cosmonaut Training Centre. It is uncertain whether or not this was a reference to his earlier appointment as Deputy Head-Political Affairs.

He is married, and he and his wife, Lilia Vladimirovna, a graduate of the First Moscow Medical Institute, have a son, Mikhail (Misha). Klimuk, who is probably the shortest man in the cosmonaut team, is said to be a quiet, strong-willed and persevering man, who is also modest and very astute. He has a real ability for technical sciences, and scientists were reportedly amazed by his ability to reproduce from memory very involved and intricate technical diagrams without the slightest error. Kovalyonok once described him as 'a walking encyclopedia'.

He is very active in public affairs, and is a member of the Central Committee of the CPSU and the Central Committee of the Young Communist League, having been a delegate to two Komsomol congresses. He was a member of the funeral committee of President Yuri Andropov in 1983.

Known to his friends as Petya, Klimuk has logged over 2000 hours in the air. His hobbies include athletics and fencing, and he is also the author of a book. He is a graduate of the V.I. Lenin Military Political Academy.

In April 1985, he reported that he was 'preparing for new space missions.'

Pyotr Klimuk was the 28th Soviet cosmonaut, and he has spent a total of 78 days 18 hours and 19 minutes in space.

Colonel Eberhard Köllner

NO PHOTOGRAPH AVAILABLE

Eberhard Köllner was born on 29 September 1939, in Strassfurt, in the German Democratic Republic.

He studied at the Franz Mehring Air Force School for Pilots, and graduated from the Gagarin Air Force Academy in 1970. He has logged over 1000 hours in jets.

Following the agreement to fly non-Soviet cosmonauts from member countries of the Intercosmos programme onboard Soyuz and Salyut spacecraft, the German Democratic Republic began a cosmonaut selection programme. The two finalists were Köllner and Sigmund Jähn and they were sent to begin their training at the Yuri Gagarin Cosmonaut Training Centre in December 1976. Jähn eventually flew the mission as research-cosmonaut onboard Soyuz 31, launched on 26 August 1978, while Köllner served as back-up, and as CapCom in FCC, using the callsign Terek-2.

Following the completion of the mission, he was awarded the German Democratic Republic's highest military award, the Scharnhorst medal, and was promoted from Lieutenant-Colonel to Colonel. He is a member of the Socialist Unity Party.

Eberhard Köllner is married, and has two children, born in 1965 or 1966, and in 1972 or 1973.

Colonel-Engineer Vladimir Mikhailovich Komarov (1927-1967)

Vladimir Komarov was born on 16 March 1927 in Moscow, the son of an intellectual. He went to the Moscow Air Force School at the age of 15, but graduated as a fighter-pilot too late to see action in World War 2. He continued his training at the Third Sassov Air Force School, at the Chkalov School in Borisoglebsk, and the Serov Flying School in Bataisk. From 1949-54, he served in various Air Force units as a fighter-pilot, and became a Test-Pilot 3rd class. He joined the CPSU in 1952, and enrolled at the Zhukovsky Military Air Academy in 1954, graduating in 1959. He was then assigned to the Air Force Research Institute.

He was selected to join the cosmonaut team in 1960, but shortly afterwards fell ill and had to go into hospital for an operation. His ill-health raised grave doubts about his ability to continue with the strenuous training programme. However, six months later, he was able to convince medical examiners of his fitness, and re-joined the training group. In another five months he is said to have caught up with his fellow cosmonauts.

It is believed that his first assignment may have been as support crew for Vostok 2, launched on 6 August 1961.

He was then chosen as the back-up to Pavel Popovich the pilot of Vostok 4, launched on 12 August 1962, and also served as the support crew for Vostok 5, launched on 14 June 1963.

Having successfully overcome one set of health problems, he soon ran into another. During training on a centrifuge, doctors were disturbed to discover that Komarov had an extra systole in his heart, leading to an occasionally irregular heartbeat – a condition known as a systolic heart murmur. He was immediately removed from the cosmonaut team. (This medical condition was extremely similar to the one that grounded the American astronaut Deke Slayton.) Komarov didn't give up, and went to all the leading heart specialists in Moscow, and accumulated a wealth of medical endorsements stating that his irregular heartbeat would cause no problems during a spaceflight. When he presented these, he was re-admitted to the cosmonaut team.

His first spaceflight came as commander of Voskhod 1, the three-man spacecraft launched on 12 October 1964. During the flight, Konstantin Feoktistov and Boris Yegorov carried out many medical and geo-physical experiments, while Komarov tested the new spacecraft flight systems. They returned to Earth after a flight lasting just 1 day and 17 minutes.

Upon his return, Komarov was awarded the title of Hero of the Soviet Union and also received the Order of Lenin.

In 1966, he was made Chief of the Engineering Department at the Yuri Gagarin Cosmonaut Training Centre. He made a second spaceflight as pilot of the one-man Soyuz 1, launched on 23 April 1967. The mission was intended as a test of the new Soyuz spacecraft and its systems, and he returned to Earth after a flight lasting 1 day 2 hours and 48 minutes. Unfortunately, the main parachute apparently became entangled, and as there was no reserve parachute, the capsule plunged to Earth, crashing into hilly country outside the town of Orenburg, killing Komarov instantly.

He was posthumously awarded the title of Hero of the Soviet Union and also received the Order of Lenin, both for the second time. In his honour, the Soviets named one of the space tracking vessels the 'Cosmonaut Vladimir Komarov', and erected a memorial to mark the spot where the Soyuz crashed.

There is some doubt that the Soviets released the full story behind the tragedy. The communication sessions between Komarov and FCC were monitored at the USAF base outside Instanbul, and these indicated serious problems with the spacecraft's attitude control during the second half of the flight. The listening post was manned by technicians of the National Security Agency, and one of these later revealed what had been overheard.

Initially, the spacecraft was rolling out of alignment every so often. After five hours of trying every mechanical possibility, the spinning got worse, and Komarov complained of feeling sick. He repeatedly asked how long it would be before re-entry. His wife was brought to FCC to talk to him and he eventually had to bid her an emotional farewell. He tried new attempts at stabilising the spacecraft, following instructions from FCC, but these failed to work. Then Premier Kosygin asked to speak to Komarov, and told him that his country was proud of him, and would always remember him. His reply was inaudible, and several minutes later, re-entry took place. His cries for help ended abruptly. When the descent module reached Earth, his charred remains were found inside. This story of course conflicts with the official version of parachute failure.

Further controversy surrounds the nature of the Soyuz 1 mission itself. It is now clear, despite official Soviet denials at the time, that a second spacecraft, Soyuz 2, was to have been launched on 24 April. Following docking, an exchange of crews would have taken place.

The Soyuz 2 prime and back-up crews have since been identified by Western observers. The prime crew were Valery Bykovsky, Alexei Yeliseyev and Yevgeny Khrunov, and the back-up crew were Pavel Belyayev/Andrian Nikolayev, Valery Kubasov and Viktor Gorbatko. However, following the initial problems with Soyuz 1, the Soyuz 2 launch was cancelled.

An unmanned Soyuz 2 subsequently flew in a rendezvous operation with Soyuz 3, piloted by Georgi Beregovoi, and in January 1969, nearly two years late, the mission intended for Soyuz 1/2 was carried out by the Soyuz 4/5 spacecraft. The prime and back-up crews of Soyuz 2 were given the same assignments on Soyuz 5 – only the commanders had changed.

Another mystery concerns training pictures released after Komarov's death, which showed him training in the same type of spacesuit as Khrunov and Yeliseyev. This would indicate that he had trained for an EVA of some description in connection with Soyuz 1/2. But there is no apparent reason for him to have conducted such an operation. During the crew transfer, he would have sealed off the descent module, while Khrunov and Yeliseyev entered the orbital module, and he would therefore not have needed a spacesuit. And on the duplicate Soyuz 4/5 mission, two years later, Vladimir Shatalov did not have a spacesuit.

Therefore, it raises the question – did Soyuz 1 only have a descent module i.e. no orbital module? If that was the case, then when Yeliseyev and Khrunov transferred to Soyuz 1 by external EVA, Komarov would have had to depressurize the descent module, hence the need for his spacesuit. Because of the possible absence of an orbital module, it has even been suggested that Soyuz 1 may in fact have been a Zond, and it is noteworthy that no pictures of Soyuz 1 have ever been released.

Komarov was the first spaceman to be killed during a spaceflight, although others from both the US and the Soviet Union had died during ground incidents. He was the first cosmonaut to make a second spaceflight, and this, together with the fact that on both occasions he was entrusted with the testing of a new spacecraft, shows the esteem in which he was held by the flight crew selectors.

He left a widow, Valentina Yakovlevna – a librarian in Star Town – and two children, Irena, born in 1958, and Yevgeny, born in 1951.

He has been described as one of the most skilled of the cosmonauts, noted in the team for his engineering knowledge and his terrific capacity for work. He served on the Council of the Union of the Supreme Soviet in the period 1962 – 66, and in his spare time, he worked as an editor at the Novosti Press Agency, contributing to, and supervising, the publication of articles and books on the Soviet space programme. Each cosmonaut apparently has a favourite star, and it is said that Komarov's was Suhail, fom the Vela Constellation.

Vladimir Komarov was the 7th Soviet cosmonaut, and he spent a total of 2 days 3 hours and 05 minutes in space.

Colonel Vladimir Vasilyevich Kovalyonok

Vladimir Kovalyonok was born on 3 March 1942, in the village of Byeloye, Krupsky District, near Minsk in Byellorussia, and his father was a farmer. After studying at the Balashov Higher Air Force School, and graduating in 1963, he served in the Soviet Air Force flying military transport planes. He logged over 1600 hours in the air, and also qualified as a paratroop instructor. He joined the CPSU in 1962 at the unusually young age of 20.

In the early 1960's, during the first spaceflights, Kovalyonok was part of the search and recovery team, and was responsible for ensuring the safe return of the landing spacecraft to the cosmodrome. It is said that that was when he made up his mind to become a cosmonaut.

In 1967, he was selected to join the cosmonaut team, and reported to Star Town in May. In the same year, he graduated from the Gagarin Air Force Academy while working at the Yuri Gagarin Cosmonaut Training Centre. He subsequently took part in the ground control of Soyuz and Salyut missions, and began training for a Salyut mission in 1972.

It is believed that Kovalyonok's first assignment was as the commander of the support crew for Soyuz 17 and Soyuz 18, launched on 10 January and 5 April 1975 respectively. (see section entitled 'Salyut 4 Crewing' for more details.)

Following the launch failure of Soyuz 18, (subsequently designated the 'April 5 anomaly'), he was moved up to the back-up position for the replacement mission, termed Soyuz 18, launched on 24 May 1975. His first spaceflight came as commander of Soyuz 25, launched on 9 October 1977. When he and Valery Ryumin attempted to dock their spacecraft with the Salyut 6 space-station, they were prevented from doing so by a fault in the docking system. They were forced to make an emergency return to Earth after a mission lasting 2 days and 45 minutes.

Kovalyonok was later awarded the Order of Lenin only – the first time that a cosmonaut had not been awarded the title of Hero of the Soviet Union following a first spaceflight, indicating official dissatisfaction with the failure of the mission.

His next assignment was as back-up to Yuri Romanenko, the commander of Soyuz 26, launched on 10 December 1977, and then as back-up to Vladimir Dzhanibekov, the commander of Soyuz 27, launched on 10 January 1978. His second spaceflight came as commander of Soyuz 29, launched on 15 June 1978, in which he made a second attempt to board Salyut 6. This time, the docking was successful, and Kovalyonok and Alexander Ivanchenkov carried out a long programme of research and experiments as well as playing host to the visiting cosmonauts from Poland and the German Democratic Republic.

During their stay onboard the Salyut, Kovalyonok and Ivanchenkov made an EVA lasting 2 hours and 5 minutes. Kovalyonok stayed in the hatchway while Ivanchenkov removed samples from the Medusa experiment and replaced them with fresh materials. The two men also tested the new spacesuits.

They returned to Earth after a flight lasting 139 days 14 hours and 48 minutes – thus setting a new duration record.

Upon his return, Kovalyonok was awarded the title of Hero of the Soviet Union and also received his second Order of Lenin. In addition, he was awarded the title of Hero of the German Democratic Republic with the Order of Karl Marx, and received the Polish Gruenwald Cross, 1st Class.

It is thought that his next assignment may have been as commander of the support crew for Soyuz 35, launched on 9 April 1980 (see section entitled 'Salyut 6 Crewing' for more details.) His next spaceflight came as commander of Soyuz T-4, launched on 12 March 1981. The mission was not originally planned and was required to provide a new long-duration crew onboard Salyut 6 to enable the Mongolian and Romanian Intercosmos missions to take place, the schedule having been disrupted by the Soyuz 33 failure. Kovalyonok began joint training with his flight-engineer, Viktor Savinykh, in November 1980.

The two men spent nearly 74 days onboard Salyut 6 carrying out a series of experiments and also acting as hosts to the Mongolian and Romanian cosmonauts onboard Soyuz 39 and 40. They then returned to Earth on 26 May after a flight lasting 74 days 18 hours and 38 minutes.

Kovalyonok was subsequently awarded the title of Hero of the Soviet Union for the second time and also received his third Order of Lenin.

In June 1985, he and Bertalan Farkas took part in the ground studies at Kursk Oblast as part of the Kursk-85 experiment conducted by the Soyuz T-13 cosmonauts.

He is married, and he and his wife Nina have a daughter, Inessa, born in 1965 or 1966, and a son, Volodya. Kovalyonok is a deputy to the Supreme Soviet of Byelorussia, and is also Chairman of the Soviet All-Union Federation of Trampolining. His hobbies include photography, skiing and hunting.

Other cosmonauts have described him as 'stubbornly persistent in his work,' and Savinykh has said of him 'Although by training, he is a pilot and an experienced cosmonaut, he has a keen interest in science.'

Natalya Ryumina, speaking of her husband's reaction to the failure of Soyuz 25, also gave an insight into Kovalyonok's character: 'I think that Valery felt he was caught up in a blind alley. Kovalyonok, with whom he went on that flight, was also really put out, but he is an emotional kind of guy and can find another channel to vent his feelings.'

Vladimir Kovalyonok was the 40th Soviet cosmonaut, and he has spent a total of 216 days 10 hours and 11 minutes in space.

Valery Nikolayevich Kubasov

Valery Kubasov was born on 7 January 1935, in the small town of Vyazniki, in the Vladimir Region, north-east of Moscow. His father Nikolai Ivanovich was a shipping-line mechanic and often took Valery on trips down the Volga. His mother's name is Tatiana Ivanovna.

He completed secondary school in 1952, with a silver medal, which meant that he did not have to take the entrance examinations for the Moscow Aviation Institute, the next step in his education. He graduated in 1958, as a mechanical-engineer in aircraft construction, and went to work in Sergei Korolyov's design bureau. He was assigned to the ballistics studies department, where he became acquainted with another future cosmonaut, Georgi Grechko.

Kubasov was engaged in designing, developing and testing space vehicles, as part of the large team designing the first manned spacecraft. At the time he had no thought of going into space himself, being more interested in the technical problems to be solved. He was, he said, 'infinitely attracted not only by calculations of the flight paths of artificial earth satellites and of interplanetary stations, but, above all, by the novelty of the work. To me, everything was a discovery.'

He says that the more complicated a calculation is, the closer it is to poetry and the more it inspires him. It has been said of him 'His motto is to get to the bottom of everything himself. He likes to travel the most difficult roads independently.'

Following Gagarin's flight, Kubasov became interested in going into space himself to test in practical flight conditions the instruments and calculations he had helped to make. And following the flight of Feoktistov onboard Voskhod 1, 'the idea of going up into space gave me no peace.'

While at the design bureau, Kubasov wrote several books and papers on the theory of flying vehicles, and his works appeared in many scientific journals. One paper, entitled 'The Correction of Interplanetary Trajectory Using Impulses of Radial Heliocentric Velocity' is said to have impressed Korolyov greatly. When Korolyov met Kubasov, and found that he was keen to make a spaceflight, Korolyov 'did everything possible to help get him on the cosmonaut team.' He recommended Kubasov to the cosmonauts detachment, saying 'This young man has brains!', and noting with approval Kubasov's enthusiasm and exacting attitude to work.

In one of his speeches, Korolyov said that specialists of different kinds would soon be needed to develop new equipment in outer space. 'Be on your toes, or else you'll be too late!' he joked. A few days later, a pile of applications appeared on Korolyov's desk. Among them was one from Valery Kubasov.

He was interviewed with Konstantin Feoktistov in 1964, but was not officially selected to join the cosmonaut team until 1966, reporting for duty in Star Town on 5 September that year. During his training, he studied for the Soviet equivalent of an MSc degree in Engineering, which he earned in 1968. He also joined the CPSU in the same year.

On 23 April 1967, Vladimir Komarov was launched onboard Soyuz 1. He was brought back to Earth after just over a day in orbit, and was killed when the spacecraft's parachute tangled following re-entry. It is now clear, despite official Soviet denials at the time, that a second spacecraft, Soyuz 2, was to have been launched on 24 April. Following docking, an exchange of crews would have taken place.

The Soyuz 2 prime and back-up crews have since been identified by Western observers. The prime crew were Valery Bykovsky, Alexei Yeliseyev and Yevgeny Khrunov, and the back-up crew were Pavel Belyayev/Andrian Nikolayev, Valery Kubasov and Viktor Gorbatko. Kubasov's first assignment therefore was as the back-up to Alexei Yeliseyev, who would have made an EVA to transfer from Soyuz 2 to Soyuz 1. However, following the initial problems with Soyuz 1, the Soyuz 2 launch was cancelled.

An unmanned Soyuz 2 subsequently flew in a rendezvous operation with Soyuz 3, piloted by Georgi Beregovoi, and in January 1969, nearly two years late, the mission intended for Soyuz 1/2 was carried out by the Soyuz 4/5 spacecraft. The prime and back-up crews of Soyuz 2 were given the same assignments on Soyuz 5 – only the commanders had changed. Kubasov therefore again served as the back-up to Alexei Yeliseyev, flight-engineer onboard Soyuz 5, launched on 15 January 1969.

In 1984, he confirmed that he had trained with Gorbatko and Gagarin in the period August 1966 – April 1967 for a flight onboard 'the new Soyuz spaceship, which was a promising but "untamed" vehicle . . .'

According to one Western report, a heavy Zond-Proton countdown was conducted on 3 and 4 July 1969, and the cosmonaut crew present at Baikonur was thought to consist of Filipchenko, Kubasov and Shonin. Their launch was cancelled when the G-vehicle failed shortly after lift-off. (see section entitled 'The Zond programme' for more details.)

The report also stated that Kubasov was scheduled to fly on the first Earth-orbital test flight of the Soviet Lunar Module. However, it should be noted that Kubasov was never a member of the lunar training group, and it is difficult to see when he could have found time to prepare for such a mission, as he was working continously on Soyuz assignments right from the time he was selected.

His first spaceflight came as flight-engineer onboard Soyuz 6, launched on 11 October 1969. During the flight, he carried out the world's first space-welding and metals smelting experiments. The Vulkan equipment was located in the orbital module of the Soyuz, and was remotely controlled by Kubasov from the descent module. In addition, Soyuz 6 carried out joint manoeuvering experiments with the Soyuz 7 and 8 spacecraft. The Soyuz 6 crew of Kubasov and Shonin returned to Earth after a flight lasting 4 days 22 hours and 43 minutes.

Upon his return, Kubasov was awarded the title of Hero of the Soviet Union and also received the Order of Lenin. He resumed work in the design bureau and continued his cosmonaut training. *There is some evidence to suggest that his next assignment was as the flight-engineer on the support crew for Soyuz 9, launched on 1 June 1970.*

He and Alexei Leonov were then assigned to the Salyut 1 programme, and the two men trained for a mission onboard the space-station, but were stood down when Kubasov became ill a few weeks before launch.

The doctors discovered he had a lung condition and removed him from the cosmonaut team. 'The months that dragged out agonisingly after they struck me off the list of cosmonauts were the hardest ones in my life,' he later recalled. 'I felt infinitely sorry about all the work put in by the big team of specialists who had trained me for the flight . . . They had plans . . . I had undoubtedly postponed them. This perhaps was what made me fight my illness. I did not believe in it. I felt it was temporary and that I could overcome it.'

He eventually recovered, and following a scrupulous examination, was placed back on the active list. *It is thought that he and Leonov were then given a second chance, and should have flown a month-long Soyuz 12 mission. However, this was cancelled following the Soyuz 11 tragedy. (See section entitled 'Salyut 1 Crewing' for more details.)*

In March 1972, a Western news agency reported that Kubasov was going to make a flight onboard a new Salyut in May of that year. The report is a strange one, particularly as it indicated that he was intended to fly with one of two rookie cosmonauts – Vinogradov or Voronov. (See section entitled 'Salyut 2 and Cosmos 557 Crewing' for more details.)

It is believed that Kubasov and Leonov were the prime crew for a mission to the Cosmos 557 Salyut, launched in May 1973. According to one Western report, shortly after the Cosmos 557 failure, the planned Soyuz 12 launch was cancelled, and Kubasov was named as flight-engineer on the Soviet spacecraft for the ASTP mission. (See section entitled 'Salyut 2 and Cosmos 557 Crewing' for more details.)

From May 1973 he trained for the ASTP mission in both the United States and the Soviet Union. If anything went wrong on the flight, the Soviets joked, Kubasov could weld it together again! (a reference to his experience on Soyuz 6.) The Soyuz 19 ASTP spacecraft was launched on 15 July 1975, and the crew of Leonov and Kubasov spent two days in orbit preparing for the historic docking with the American Apollo spacecraft. The two craft remained docked for nearly two days, during which the crews carried out joint experiments and transfers to each others spacecraft. Then the two craft separated, and the Soyuz 19 crew returned to Earth after a flight lasting 5 days 22 hours and 31 minutes.

Upon his return, Kubasov was awarded the title of Hero of the Soviet Union and also received the Order of Lenin, both for the second time.

His next assignment came as the back-up to Pyotr Klimuk, the commander of Soyuz 30, launched on 27 June 1978. *He may originally have been the prime commander for the mission, but was switched to the back-up crew shortly before launch. (See section entitled 'Intercosmos Crewing' for more details.)* In December 1978, he was teamed up with Bertalan Farkas, to train for the joint Soviet-Hungarian spaceflight. Whilst in training, he was involved in the flight-control of Soyuz 32/Salyut 6.

Kubasov and Farkas were launched onboard Soyuz 36 on 26 May 1980. They linked up with Salyut 6, and joined the crew of Popov and Ryumin already onboard. Kubasov and Farkas spent nearly seven days onboard carrying out a programme of joint Soviet-Hungarian experiments. They then returned to Earth after a flight lasting 7 days 20 hours and 46 minutes. Upon his return, Kubasov was awarded the Order of Lenin for the third time, together with the Gold Star of Hero of the Hungarian Peoples Republic.

He is married, and his wife Lyudmila Ivanovna, an aeronautical engineer and a graduate of the Moscow Aviation Institute, works in a scientific research institute. They have a son Dimitry (Dima), born in 1970 or 1971 and a daughter Ekaterina (Katya), born in 1965 or 1966. Kubasov is a sports enthusiast, and a fine athlete, enjoying skiing, water-skiing and hunting. He has been interested in fishing and camping since childhood, and is also a keen photographer, ranking film-making as one of his main hobbies. He is the author of a book entitled ' Touch Space ', published in 1984, and is Vice-President of the Soviet-American Friendship Society, and co-chairman of the Spaceflight section of the IAF. One of his oft-repeated ambitions is to fly to Mars.

Kubasov is a stocky man, rather shy and quiet, held in high regard by his friends and colleagues, who describe his knowledge of spacecraft design as 'encyclopaediac'. He is known amongst his fellow cosmonauts as 'The Brain' because of his vast knowledge of engineering, ballistics, astronomy and planetary physics. Alexei Leonov once described him as an 'absolutely first-class engineer', while Bertalan Farkas said 'he is a calm and pleasant person . . . a very hard-working and experienced cosmonaut . . . always ready to help'.

Valery Kubasov has received many awards and honours apart from those already listed. They include the Tsiolkovsky Gold Medal from the Soviet Academy of Sciences; the medal 'For valiant labour in commemoration of V.I. Lenin's centenary'; the emblem 'for active participation in Komsomol activities'; the Gold Medal of the Yugoslavian Academy of Sciences; the medal 'for development of virgin lands'; and a number of medals from other countries. In addition, he is an honorary citizen of Karaganda, Kaluga, Vladimir, Vyazniki and Houston.

Valery Kubasov was the 18th Soviet cosmonaut, and he has spent a total of 18 days 18 hours and 00 minutes in space.

Colonel Vasily Grigorievich Lazarev

Vasily Lazarev was born on 23 February 1928, in the village of Poroshino in the Altai region of Southern Siberia. He grew up in Sverdlovsk in the Urals, where his parents still live. He attended the Sverdlovsk Medical Institute, and then enrolled at the Saratov Medical Institute – a military school. He graduated with a medical degree in 1952, specialising in aviation medicine. He then began work as an Air Force doctor, with the rank of Lieutenant, and enrolled at the Chuguyev Air Force School, spending two years learning to fly high performance jets. From 1954 onwards, he was a flight-instructor, test-pilot, and flight equipment tester in the Soviet Air Force.

In 1954, Lazarev married an aerospace engineer, and in 1956 he joined the CPSU. He applied to join the cosmonaut team in 1960, but was turned down following a detailed physical examination, as he didn't meet the exacting standards required. He returned to his work as an Air Force flight-surgeon, specializing in aviation medicine and crew systems.

He did however serve as back-up to Boris Yegorov, the flight-surgeon onboard Voskhod 1, launched on 12 October 1964, and following the completion of that mission, he was stood down. He was then officially selected to join the cosmonaut team in 1966, together with Oleg Makarov who had also been on the Voskhod 1 back-up crew.

His first assignment following his official selection was as back-up to Andrian Nikolayev, the commander of Soyuz 9, launched on 1 June 1970. *It is thought that Lazarev may have been intended to fly to the Cosmos 557 Salyut launched in May 1973. (See section entitled 'Salyut 2/Cosmos 557 Crewing' for more details.)* His first spaceflight came as commander of Soyuz 12, launched on 27 September 1973. During the flight, which was the first Soviet mission since the Soyuz 11 deaths, Lazarev and Oleg Makarov tested improved spacecraft systems, carried out Earth observations, and tested a new lightweight spacesuit. They returned to Earth after a flight lasting 1 day 23 hours and 16 minutes.

Upon his return, Lazarev was awarded the title of Hero of the Soviet Union and also received the Order of Lenin.

His next assignment was as commander of what should have been Soyuz 18, launched on 5 April 1975. Lazarev and Oleg Makarov's flight was intended to be a 60 day mission to

Salyut 4. However, 4 minutes and 51 seconds after lift-off, a booster malfunction caused the automatic abort sequence to begin.

The launch escape system pulled the Soyuz Descent Module clear from the booster, and it then made an emergency touchdown, landing in the snow-covered Altai Mountains. The capsule's parachute apparently caught in a tree right over a gorge, at the end of a mission lasting only 22 minutes. Upon his return, Lazarev was awarded the Order of Lenin, for 'courage displayed during the launch.'

The aborted mission was subsequently designated 'the April 5 anomaly', and the back-up crew flew the replacement mission – Soyuz 18.

Following the failure of the mission, it appears that Lazarev was taken off the active list, and he was appointed Training Manager at the Yuri Gagarin Cosmonaut Training Centre in 1975, replacing Valery Bykovsky. In October, 1976, he became responsible for the training of the Intercosmos cosmonauts.

At some point however, Lazarev was placed back on the active list, and he served as the back-up to Leonid Kizim, the commander of Soyuz T-3, launched on 27 November 1980. As he has not subsequently made a spaceflight, it is assumed that he remains Training Manager for the cosmonaut team.

In August 1982, he took part in the Soyuz T-7 post-flight press conference, probably in his capacity as Training Manager.

He is married, and he and his wife Luiza, who works at the Yuri Gagarin Cosmonaut Training Centre, have a son, Alexander – a pilot in the Soviet Air Force. Lazarev is Vice President of the NDRI Friendship Society, and has written a biography of Vladimir Komarov entitled 'Ispetatyel Kosmicheskikh Korablyee' published in 1979.

Vasily Lazarev was the 26th Soviet cosmonaut, and he has spent a total of 1 day 23 hours and 38 minutes in space.

Valentin Vitalyevich Lebedev

Valentin Lebedev was born on 14 April 1942, in Moscow. He enrolled at the Moscow Aviation Institute in 1960, and it said that when he learnt of the Vostok 1 flight, he wrote a letter to Yuri Gagarin asking to be put on the cosmonauts list. In his free time, he attended an aviation club and piloted aircraft, helicopters and gliders. He graduated from the Institute in 1966 with the equivalent of a MSc degree in aviation engineering. He then went to work in Sergei Korolyov's design bureau as an engineer, and whilst there, met his future wife. He took an active part in the development and testing of new spacecraft systems, and proved himself to be a knowledgeable and imaginative specialist.

He joined the CPSU in 1971, and was specially selected to join the cosmonaut team in 1972, being the only person chosen that year. He has stated that his first assignment was ASTP related, and he was placed in a training group which eventually flew the Soyuz 13, 16, 19 and 22 missions. His first spaceflight came as flight-engineer onboard Soyuz 13, launched on 18 December 1973. During the flight, Lebedev and Pyotr Klimuk carried out studies of the stars using the Orion 2 telescope array, for which Lebedev was responsible. It was said that to successfully accomplish all his assigned tasks, he had to master the skills of an astronomer, geologist and botanist. The two men also carried out medical experiments and tested new spacecraft systems before returning to Earth after 7 days 20 hours and 56 minutes in space.

Upon his return, Lebedev was awarded the title of Hero of the Soviet Union and also received the Order of Lenin. *There are some puzzling aspects to the Soyuz 13 crewing, including the fact that Lebedev had only been a cosmonaut for one year before flying, and did not perform any back-up duties. In addition, he is known to have begun his training with Vladimir Kovalyonok, whilst Klimuk was teamed with Vitaly Sevastyanov in early 1972, and later flew a long-duration mission with him in 1975. It is therefore possible to speculate that Lebedev was originally the back-up and that there was a last-minute switch when Sevastyanov became disqualified for some reason. It is interesting to note that the Soyuz 13 back-up crew has never been identified.*

It emerged some years later that Lebedev was chosen for Soyuz 13 because he was a specialist in the long-duration solo Soyuz. The main objective of the mission was to test the Soyuz spacecraft in preparation for the forthcoming ASTP and Salyut flights. Klimuk was Salyut-trained, and was a specialist in the Orion 2.

The Orion 2 studies yielded good results and Lebedev later used the data in a thesis. He defended his dissertation in 1974, and was awarded a Candidate of Technical Sciences degree.

It is believed that his next assignment was as flight-engineer on the support crew for Soyuz 25, launched on 9 October 1977. He should then have served as back-up flight-engineer for the next mission, before making his second spaceflight on the next long-duration flight. However, following the failure of Soyuz 25, the crewing schedules were revised and it is believed that he lost his place to Valery Ryumin, and was given another support crew assignment, this time for Soyuz 29. (see section entitled 'Salyut 6 Crewing' for more details.)

He then served as back-up to Ryumin, who flew as flight-engineer onboard Soyuz 32, launched on 25 February 1978. He was then chosen as flight-engineer for Soyuz 35, launched on 9 April 1980, and began joint training for the mission with Leonid Popov. Unfortunately, Lebedev wrenched his knee during exercises on the trampoline at the last physical training session, some 6 weeks before the flight, and was hospitalised. It originally looked like an ordinary sprain, and it seemed that he would recover in 2-3 days.

However, it soon became clear that he had injured his knee ligaments and an operation was necessary. He therefore lost his place on the mission, and was replaced by Valery Ryumin, who had just returned from a mission lasting 175 days. By a strange co-incidence, it appears that Lebedev was twice prevented from flying to Salyut 6, and on both occasions, was replaced by Valery Ryumin !

After a period of convalescence, Lebedev resumed training, this time with Anatoly Berezovoi, and a year later, he made a second spaceflight as flight-engineer onboard Soyuz T-5, launched on 13 May 1982. He and Berezovoi docked their spacecraft with the orbiting Salyut 7 space-station and carried out a long programme of research and experiments as well as playing host to the joint Soviet-French spaceflight Soyuz T-6, and the visiting Soyuz T-7.

During their stay onboard the Salyut, Lebedev and Berezovoi made an EVA lasting 2 hours and 33 minutes. Berezovoi stayed in the hatchway while Lebedev removed samples from the Medusa experiment and replaced them with fresh materials. The two men also tested new tools and assembly procedures.

They returned to Earth after a flight lasting 211 days 9 hours and 5 minutes – thus setting a new duration record.

Lebedev was subsequently awarded the title of Hero of the Soviet Union and also received the Order of Lenin, both for the second time. In addition, he was awarded a 'Distinguished Prospector' badge by the Soviet Union's Ministry of Geology in recognition of his work in Earth observations during the 211 day flight. He is the only cosmonaut to be awarded this high Soviet honour. In addition, he was made a member of the French Legion of Honour.

In 1983, a lengthy article was published in *Pravda* consisting of extracts from Lebedev's diary. His frank remarks about the loneliness of living in space, away from family and friends, problems with insomnia despite a tiring workload, and establishing a working relationship with Berezovoi, revealed the psychological pressures involved in long-duration spaceflight.

Lebedev is married, and his wife Lyudmila, known as Lyusya, works as an engineer in a design bureau. They have a son, Vitaly, born in 1972 or 1973.

Lebedev is Chairman of the Soviet Federation of Acrobatics, and he also supports and acts as advisor on space-engineering to Iskra – the body of Moscow Aviation Institute student designers who have had piggyback satellites launched from Salyut space-stations.

Valentin Lebedev was the 29th Soviet cosmonaut and he has spent a total of 219 days 6 hours and 01 minute in space.

Major-General Alexei Arkhipovich Leonov

Alexei Leonov was born on 30 May 1934, in Listvyanka, in the Kemerovo Region, the eighth of a family of nine. By the age of 15, he was said to be an expert and experienced cyclist, having earlier made a bicycle from discarded parts and cycled long distances every day.

He began drawing as a boy, and in 1941, when the Siberian town where he lived started receiving trainloads of casualties from the front, he started spending hours in the hospital, drawing soldiers and officers and battle scenes.

Soon after the war, the family moved to Kaliningrad on the Baltic Coast, and Leonov switched to seascapes. His idol was the famous Soviet marine artist, Ivan Aivazovsky.

His school and his parents expected him to become a professional artist and upon finishing secondary school in Kaliningrad in 1953, he enrolled at the Academy of Arts in Riga. But almost at once, he left it and entered the Chuguyev Air Force School, graduating with an honours diploma in 1957. He then qualified as a paradrop instructor in the Soviet Air Force, making more than 100 jumps himself. While at flying school, he took up fencing, volley ball and yachting, and resumed his art studies by attending evening classes attached to the Academy of Arts in Riga. He joined the CPSU in 1957.

During the search for cosmonauts in the late 50's, the examining board told Leonov that he was under consideration for 'something new, something very, very difficult.' He was selected to join the cosmonaut team in 1960, and learnt of his selection on 5th March. It is said that because he was such an agile, athletic, and competent pilot, as well as an outstanding parachutist, he was, at one time, the firm favourite to fly as pilot of Vostok 1. He made no secret of the fact that his one ambition was to be the first man on the Moon.

In 1963, Leonov and his wife were both nearly killed when their car left the road and plunged through the ice of a frozen pond. Leonov struggled out, and rescued both his wife and the driver of the car. On another occasion, only his great strength saved him from certain death. He was making a parachute jump using the Vostok ejector seat, and at separation, one of his straps caught on the ejector seat, which meant that he would probably have landed on his head. So he struggled with the seat, eventually managing to bend the protruding duralumin metal, so freeing his parachute, enabling him to make a perfect landing. Later, the seat had to be bent back into shape using a hammer.

Whilst training for his first spaceflight, Leonov, in just one year, ran 311 miles (500km), cycled 621 miles (1000km), skied 186 miles (300km), and spent hundreds of hours on all kinds of sports equipment. Finally, after all this training, he made his first spaceflight together with Pavel Belyayev onboard Voskhod 2, launched on 18 March 1965.

During the flight, Leonov made history by becoming the first man to walk in space. However, when he tried to make his way back into the Voskhod 2 capsule, he ran into difficulty. Soviet officials maintain that he had trouble in re-entering the air-lock with a cine-camera that kept getting in his way. Western observers, however, have suggested that his spacesuit 'ballooned', causing him great difficulty in moving. In any event, Leonov apparently had to struggle for nearly 12 minutes to get back into the spacecraft, eventually succeeding after an EVA lasting 24 minutes.

At the end of the mission, Belyayev had to make a manually-controlled re-entry following the failure of the automatic system. This meant that the two men had to make an additional orbit, and they returned to Earth after a flight lasting 1 day 2 hours and 2 minutes. The Voskhod capsule touched down above the snow-line in a forest near Perm, in the Urals, some 1250 miles (2000km) from the target zone. The capsule landed between two enormous fir trees in deep snow, and when the snow gave way, the spaceship became dangerously wedged between the two trees. The two men had to wait 2½ hours for the recovery helicopter, and then had to stay overnight at the touchdown spot.

Upon his return, Leonov was awarded the title of Hero of the Soviet Union and also received the Order of Lenin. During his spaceflight, he put his artistic talents to good use and drew sketches using crayons and the ships log. Following the flight, he made many vivid and impressionistic paintings of his spacewalk and the view of the Earth. He was admitted to the Soviet Artists Union and started taking part in Art exhibitions.

According to Soviet sources, following the Soyuz 1 accident in April 1967, 'A new flight programme was drawn up, and Alexei Leonov was included in the complex stage of training . . . everyone was waiting impatiently for Spring (1968). There would be new flights and new plans . . .'

Whilst the nature of the 'new flights' was not revealed, it is known that Pavel Belyayev started training for a Zond circumlunar flight in late 1967. The flight was probably scheduled for late 1968, and according to one Western report, was cancelled at the last moment. According to the same report, Leonov was Belyayev's back-up.

It has now been confirmed that in mid-1967, Leonov was assigned to the Zond training group, and began preparing for a flight to the Moon. *(See section entitled 'The Zond programme' for more details.)*

He and Yuri Gagarin were in training together shortly before Gagarin was killed in March 1968, and later that year, Leonov graduated from the Zhukovsky Military Air Academy.

Following the abandonment of the Soviet lunar programme, he was re-assigned to the Salyut 1 programme. He and Valery Kubasov trained for a mission onboard the space-station, but were stood down when Kubasov became ill a few weeks before launch. *It is thought that following Kubasov's recovery, the two men were given a second chance, and should have flown a month-long Soyuz 12 mission. However, this was cancelled following the Soyuz 11 tragedy.* *(See section entitled 'Salyut 1 Crewing' for more details.)*

It is believed that the two men were then the prime crew for a mission to the Cosmos 557 Salyut, launched in May 1973. According to one Western report, shortly after the Cosmos 557 failure, the planned Soyuz 12 launch was cancelled, and Leonov was named as commander of the Soviet spacecraft for the ASTP mission. (See section entitled 'Salyut 2 and Cosmos 557 Crewing' for more details.)

The joint Soviet-American flight was scheduled for 15 July 1975, and upon being given the assignment, Leonov is reported to have complained 'But I don't even speak English!' Shatalov is then said to have replied 'That's alright – you have two years and two months to learn.' If correct, this would confirm that Leonov and Kubasov were to have flown a May 1973 Salyut mission.

In an interview in 1975, Leonov indirectly confirmed that he was in training to command a Salyut station **after** Salyut 1. 'Just to learn to operate the solar telescope I spent three months at the Crimea Observatory,' Leonov said. However, there was no solar telescope on Salyut 1. The first successful instrument was carried on Salyut 4, as witnessed by its designation: OST-1. This would appear to be further confirmation that Leonov was intended to fly a 1973 Salyut mission.

After being named as commander of the Soviet half of the Apollo-Soyuz Test Project in May 1973, he trained for the mission in both the United States and the Soviet Union. He was tremendously popular with the American astronauts and space officials, who found him to be a 'very funny guy', always telling jokes. But he knew when to be serious, and was reported to be an extremely hard worker.

The Soyuz 19 ASTP spacecraft was launched on 15 July 1975, and the crew of Leonov and Valery Kubasov spent two days in orbit preparing for the historic docking with the American Apollo spacecraft. Following the docking, Leonov and the Apollo commander Tom Stafford shook hands in space. The two spacecraft remained docked for nearly two days, during which the crews carried out joint experiments and transfers to each others spacecraft. During these transfers, Leonov presented each of the Apollo astronauts with a very special and unique gift. The presents were portraits of each man, as sketched by Leonov in space. The two craft then separated, and the Soyuz 19 crew returned to Earth after a flight lasting 5 days 22 hours and 31 minutes.

Upon his return, Leonov was awarded the title of Hero of the Soviet Union and also received the Order of Lenin, both for the second time. He was also promoted from Colonel to Major-General. He is married and his wife Svetlana Petrovna (or Pavlovna) is an editor, and is a graduate of a teachers institute. They married in 1960, and have two daughters, Vika (Viktoria) born in 1961, and Oksana. Leonov's hobbies include filming, hunting, water-skiing, and of course, parachute jumping.

He has written ten or more books about space, including a children's book entitled 'I Walk in Space' published in 1979, and 'Man and Universe', which he co-wrote with Sokolov, published in 1984. In 1979, he co-wrote the script of a Soviet feature film entitled 'Star System.'

Leonov's fame as an artist has spread far and wide. His paintings have been incorporated in several Soviet stamp issues, and his work has been exhibited in Moscow, Orel, Simferopol, Bratislava, Prague, Ottowa, Helsinki, and in the Bjenal annual show. He is a member of the Artists Union of the Soviet Union, and during his long cosmonaut training, he put his artistic talents to good use, and designed, wrote and illustrated the one-page cosmonaut newspaper 'Neptune', of which he was appointed Editor-in-chief.

'Why do I Paint ?' he once asked. ' Because I want to. I am looking for a means of self-expression. The pencil and brush help me to understand man, myself included, to reveal the spiritual world of my contemporaries, to delve into their thoughts and feelings, to make man better . . . It was aviation that made me an artist'

Leonov is a keep-fit enthusiast, and during his long training for ASTP, he entered for a Soviet physical fitness award scheme designed for people in the 25-30 age group. However, Leonov, at the age of 41, managed to meet all the requirements, and was awarded a gold badge. He had to train for quite some time to knock 4 seconds off his time for swimming the breaststroke. He is actually chairman of the national council for this award, known as the Ready for Labour and Defence award scheme. Apparently, before appointing him as commander of the Soviet ASTP crew, the doctors said that he had to lose 11 lbs (5 kg), as he was overweight. He therefore made his training programme even tougher, and trained for 4 hours a day.

He is Vice-President of the Soviet-Italian Friendship Society, and deputy-chairman of the Soviet-Egyptian Friendship Society, as well as a deputy to the Moscow regional Soviet, and a member of the Central Committee of the Young Communist League (YCL). He is also the Chairman of the Constituent Council of the Press and News Agency.

Leonov has received many awards and honours apart from those already listed. These include the title of Hero of the DRV, Hero of the Republic of Bulgaria, the Tsiolkovsky Gold Medal, the Great Gold Medal of the FAI, 1st degree diploma, the Great Gold Medal of the Czechoslovakian Academy of Sciences (for Services to Mankind), the Gold Damask blade, the 1977 Harmon International Aviation Trophy (together with Tom Stafford), and a total of 18 orders and medals, including 8 from other countries.

He is also an honorary citizen of the towns of Kemerovo, Kaliningrad, Vologda, Kremenchug, Nalchik, Belgorod, Drogobych, Sochi, Chuguyev, Termez, Kapsukas, Perm, Sofia (Bulgaria), Ust-na-Labe (Czechoslovakia), Altenburg (GDR), and Houston (USA).

He is presently Deputy Head of the Yuri Gagarin Cosmonaut Training Centre, and head of the international detachment. Of this position, Leonov said during a pre-ASTP interview ' I am Deputy Head of the Gagarin Cosmonaut Training Centre. This means that my job is to train cosmonauts, test them and recommend them for flights. In order to have the moral right to do so, one must be able to perform all this oneself, at least as well as the people one is testing. I have been an instructor of the Vostok, Voskhod and Soyuz spaceships. In the years which I have devoted to the profession of cosmonaut I have taken part in almost all the Soviet projects. Sad though it may be, I am now a space veteran.'

Sergei Korolyov had high praise for Leonov as a cheerful, sturdily-built pilot with a rare gift for getting along with people, and as a man who literally bubbled with energy. 'In my opinion,' he said, 'Alexei Leonov's main trait is liveliness of mind. That is the first thing. The second is his good technical knowledge and the third is his fine character. He is an artist, he paints pictures himself, and he is also a very sociable and, to my mind, good natured and likeable person. He is a bold pilot and can handle modern jet fighters with great technical proficiency. I believe this man deserves the greatest trust.'

One of Leonov's biographers described him as a 'sun-kissed' man – because of ' his sandy-coloured hair, his mobile, freckled face with mischievous, sparkling eyes, and his warm sincere relations with people.' Yuri Gagarin said of him 'Such men can be found one in ten thousand. He's a jack of all trades. It's a pleasure to work with such a person.'

He has also been described as 'A lively, energetic, cheerful, witty and happy-go-lucky fellow,' although he is said to be moody at times. Others have described him as a man of endurance, cold-bloodedness, high discipline and magnificent physical development, '. . . volatile, poetic, and always ready to take a risk,' with a high capacity for work, an unfailing interest in his work, optimism and courage, persistence and patience.

In September 1976, he was said to be closely involved with the Soyuz 22 mission, and in June 1977, he was pictured with the Soyuz 24 crew, and may have been involved in their training. In November 1983, he accompanied Lyakhov and Alexandrov on their flight back to Baikonur following the completion of the Soyuz T-9 mission.

Alexei Leonov was the 11th Soviet cosmonaut, and he has spent a total of 7 days and 33 minutes in space.

Captain Bui Thanh Liem (1949-1981)

NO PHOTOGRAPH AVAILABLE

Bui Thanh Liem was born on 30 June 1949, in Hanoi, North Vietnam, and his father was killed in the Vietnam war when he was just one year old. He was bought up by his mother, Nguyen Ngoc Thi Yen, who worked at the Thong Nhat Match factory. She is now Head of the Personnel Department of the Hai Ha Food Factory.

As an only son of a war hero, Liem could be exempted from military service, but his mother encouraged him to enlist in 1966, when he came of age. He entered the Air Force and was sent to the Soviet Union for technical training, which he is said to have completed with outstanding grades in November 1970. Upon his return to North Vietnam he was made a Sub-Lieutenant and was assigned as an interceptor pilot with the 'Red-Star' Regiment. He was credited with the downing of two US planes, and on one occasion in 1972, together with four other pilots, is said to have shot down five US F4 Phantoms.

He joined the Vietnamese Communist Party in 1973, and was promoted to First-Lieutenant in 1974. He was then sent to the Gagarin Air Force Academy in the Soviet Union, and returned to North Vietnam in 1978, when he was promoted to Captain and given a staff position.

In May 1979, Vietnam became a member of the Intercosmos programme, and as such was immediately invited by the Soviet Union to participate in a joint spaceflight, in the same way as the other member countries of Intercosmos. It was then revealed that Vietnamese cosmonauts were already undergoing training at the Yuri Gagarin Cosmonaut Training Centre.

Liem and Pham Tuan were the two cosmonauts, and they had begun their training in April 1979. Liem was paired with Valery Bykovsky in September 1979, and Tuan eventually flew as research-cosmonaut onboard Soyuz 37, launched on 23 July 1980. Liem probably served as CapCom at FCC.

On 30 September 1981, the East German newspaper 'Neves Deutschland' reported that Hanoi had announced the death of Bui Thanh Liem, in an air crash. No other details were given.

The holder of The Order of Glorious Combatant 2nd and 3rd class, he was married with a daughter, born in 1976 or 1977.

Captain José Armando Lopez-Falcon

NO PHOTOGRAPH AVAILABLE

José Lopez-Falcon was born on 8 February 1950, in Havana, Cuba and became a pilot at the age of 17.

Following the agreement to fly non-Soviet cosmonauts from member countries of the Intercosmos programme onboard Soyuz and Salyut spacecraft, Cuba began a cosmonaut selection programme. A large number of Cuban Air Force pilots were chosen for the aptitude tests and medical examinations. Nine finalists were chosen, of which four were sent to the Soviet Union in January 1978. They spent the last ten days of January and all of February undergoing numerous tests before returning to Cuba.

The two finalists were Lopez-Falcon and Arnaldo Tamayo-Méndez, who learnt of their selection on 1 March 1978 and began their training at the Yuri Gagarin Cosmonaut Training Centre on 22 March 1978. Lopez-Falcon joined the Communist Party of Cuba in 1979, and teamed up with Yevgeny Khrunov in September 1979. Tamayo-Méndez eventually flew the mission as research-cosmonaut onboard Soyuz 38, launched on 18 September 1980, while Lopez-Falcon served as back-up, and probably as CapCom in FCC.

In Spanish-speaking countries, it is the normal practice for a child to take its mother's family name as the last name. Nonetheless, it is the father's name, placed next to last, which is regarded as the surname, or family name, under which entry is made in alphabetical lists. Therefore, José Armando Lopez-Falcon should be referred to as Lopez, or Lopez-Falcon, but never as Falcon.

He is married, and has one child.

Colonel Vladimir Afanasevich Lyakhov

Vladimir Lyakhov was born on 20 July 1941, in Antratsit, in the Voroshilovgrad region of the Ukraine, in a large coalmining area. His father, a miner, was killed during World War Two when Lyakhov was only 3, and from the age of 7, he was brought up by his step-father, Mikhail Grigoryevich Yurov. Upon leaving school, Lyakhov enrolled in the Kharkov Higher Air Force School, and joined the CPSU in 1963. Upon graduation in 1964, he joined the Soviet Air Force and served in a fighter squadron in the Soviet Far East, where he became an experienced pilot. He flew in 14 different types of aircraft, including MiG-21's, and logged 1300 hours flying time.

Lyakhov was selected to join the cosmonaut team in 1967, and reported for duty in May of that year. Shortly afterwards he qualified as a test-pilot, and holds the titles 'Military Pilot, 1st Class' and 'Test pilot, 3rd Class.' In 1975, he graduated from the Gagarin Air Force Academy while continuing his work at the Yuri Gagarin Cosmonaut Training Centre. In the same year he was assigned to a group of cosmonauts selected to prepare for flights to Salyut space stations.

It is believed that his first assignment was as commander of the support crew for Soyuz 25, launched on 9 October 1977. He should then have served as back-up commander for the next mission, before making his first spaceflight as commander of the next long-duration flight. However, following the failure of Soyuz 25, it is believed that he served as commander of the support crew for Soyuz 26, before moving up to become back-up commander for the next mission. (see section entitled 'Salyut 6 Crewing' for more details.)

His next assignment was therefore as back-up to Vladimir Kovalyonok, the commander of Soyuz 29, launched on 15 June 1978. His first spaceflight came as commander of Soyuz 32, launched on 25 February 1979. He and Valery Ryumin docked their spacecraft with the orbiting Salyut 6 space-station and carried out a long programme of research and experiments. They were also intended to act as hosts to the visiting joint Soviet-Bulgarian flight, Soyuz 33, before its engine malfunctioned, causing the mission to be aborted.

When a radio-telescope attached to the Salyut became entangled with other equipment, Lyakhov and Ryumin had to carry out an EVA lasting 1 hour and 23 minutes, during which they successfully cut the telescope free. The two men returned to Earth after a flight lasting 175 days and 36 minutes.

Upon his return, Lyakhov was awarded the title of Hero of the Soviet Union and also received the Order of Lenin. During the long stay onboard Salyut 6 he was also awarded the title of 'Outstanding Worker in the Fishing Industry' by the Soviet Marine Agency, because of his work in locating plankton fields in the oceans where large shoals of fish had congregated.

During the mission, Lyakhov had celebrated his 38th birthday. It was also revealed that he had been a smoker for 28 years and had had to give up just before the mission.

His next assignment was as back-up to Vladimir Dzhanibekov, the commander of Soyuz 39, launched on 22 March 1981. *It is believed that he then served as commander of the support crew for Soyuz T-5, launched on 13 May 1982, following which he almost certainly served as the back-up to Vladimir Titov, the commander of Soyuz T-8, launched on 20 April 1983. (see section entitled 'Salyut 7 Crewing' for more details.)*

His second spaceflight came as commander of Soyuz T-9, launched on 27 June 1983. He and Alexander Alexandrov docked with the orbiting Salyut 7 space-station, and settled down to a long programme of scientific experiments and observations.

A problem arose with the Salyut's solar panels, resulting in a reduced flow of electrical power. This seriously affected the environmental control system, resulting in cold, damp and uncomfortable conditions which threatened to make the station uninhabitable, and also posed a threat to the continued function of the station's electronics and other systems.

To overcome this serious problem, a repair crew of Vladimir Titov and Gennady Strekalov were assigned to train for a mission launched specifically to repair the faulty Salyut 7 solar array. However, following a docking failure, and then an aborted launch, Lyakhov and Alexandrov, who had not been trained for such an operation, had to complete the repair themselves, by using detailed instructions read from the ground.

The repairs involved two separate EVA's, of 2 hours 50 minutes and then 2 hours 55 minutes, during which the two men had to fit two new solar panels sent up by a Progress spacecraft. Lyakhov became the first cosmonaut to complete two EVA's, having made his first from Salyut 6. The repairs were successful, and the crew resumed their normal programme. They returned to Earth after a flight lasting 149 days 10 hours and 46 minutes.

The Soyuz T-9 mission was originally planned to last approximately 100 days, but following Titov and Strekalov's launch abort, it was extended by 50 days.

Lyakhov was subsequently awarded the title of Hero of the Soviet Union and also received the Order of Lenin, both for the second time.

He is married, and he and his wife, Zinaida, an accountant/economist, have a son, Yuri, and also a daughter. Lyakhov is a passionate hockey fan, and is said to be 'self-restrained' in character. He is Vice-President of the Soviet-Sri Lanka Friendship Society.

Vladimir Lyakhov was the 45th Soviet cosmonaut, and he has spent a total of 324 days 11 hours and 22 minutes in space.

Major Bela Magyari

Bela Magyari was born in August 1949, in Kiskunfelegyhaza in Southern Hungary. After completing his secondary studies, he became a student at the Gyorgy Kilian Aeronautical Engineering College in Szolnok in 1967. Upon graduation in 1969, he went to the Soviet Union to continue his studies.

He was made a Flight-Lieutenant in 1972, and was assigned to the Hungarian Air Force, subsequently being promoted to Captain (Squadron Leader) in 1978. Magyari was awarded the title of Fighter-pilot 3rd class in 1973, and promoted to first-class in 1977. He joined the Hungarian Socialist Workers Party in 1974, and has been decorated on four occasions.

Following the agreement to fly non-Soviet cosmonauts from member countries of the Intercosmos programme onboard Soyuz and Salyut spacecraft, Hungary began a cosmonaut selection programme. The two finalists were Magyari and Bertalan Farkas and they were sent to begin their training at the Yuri Gagarin Cosmonaut Training Centre, arriving on 20 March 1978. Following nine months of general training, Magyari was teamed up with Vladimir Dzhanibekov in December 1978. Farkas eventually flew the mission as research-cosmonaut onboard Soyuz 36, launched on 26 May 1980, while Magyari served as back-up, and also as CapCom in FCC.

Following the successful completion of the mission, Magyari was promoted from Captain to Major, in recognition of his 'outstanding preparations for the joint Soviet-Hungarian spaceflight.' He received many awards and honours including:- The Order of the Banner of the Hungarian Peoples Republic, adorned with Laurel Wreath; the Golden Wreath badge of the Hungarian-Soviet Friendship Society; and the title of Cosmonaut of the Hungarian Peoples Republic.

Bela Magyari is married, and his wife, formerly a glider pilot, is now an office worker. They have a daughter, Greta, born in 1975 or 1976. Magyari's father is a school caretaker, and his mother is an administrator.

Oleg Grigoryevich Makarov

Oleg Makarov was born on 6 January 1933, in the village of Udomlya, in the Udomel District of Kalinin Region. He comes from an Army family, and his parents now live near Leningrad. In 1951, he went to the E.N. Bauman Moscow Higher Technical School, graduating in 1957 with a degree in engineering. He then went to work in Sergei Korolyov's design bureau as an engineer. He took part in developing the control panel of the Vostok spacecraft, as well as being involved in the design of the Voskhod and Soyuz spacecraft. He joined the CPSU in 1961.

He gave lectures to the cosmonauts on spacecraft construction, and helped prepare the flights of Vostok 1 and Vostok 2. He then served as back-up to Konstantin Feoktistov, the researcher onboard Voskhod 1, launched on 12 October 1964, and following the completion of that mission, he was stood down. He was then officially selected to join the cosmonaut team in 1966, and arrived in Star Town in November of that year.

In mid-1967, he was assigned to the Zond training group, and began preparing for a flight to the Moon. *According to one report, Makarov was the Chief Engineer for the Soviet Lunar Module, and was scheduled to fly on the first lunar landing mission. (see section entitled 'The Zond programme' for more details.)* Upon the abandonment of the Soviet lunar programme, he was re-assigned to the Soyuz programme.

It is thought that Makarov may have been intended to fly to the Cosmos 557 Salyut launched in May 1973. (See section entitled 'Salyut 2/Cosmos 557 Crewing' for more details.) His first spaceflight came as flight-engineer onboard Soyuz 12, launched on 27 September 1973. During the flight, which was the first Soviet mission since the Soyuz 11 deaths, Makarov and Vasily Lazarev tested improved spacecraft systems, carried out Earth observations, and tested a new lightweight spacesuit. They returned to Earth after a flight lasting 1 day 23 hours and 16 minutes.

Upon his return, Makarov was awarded the title of Hero of the Soviet Union and also received the Order of Lenin.

His next assignment was as flight-engineer onboard what should have been Soyuz 18, launched on 5 April 1975. Makarov and Vasily Lazarev's flight was intended to be a 60 day mission to Salyut 4. However, 4 minutes and 51 seconds after lift-off, a booster malfunction caused the automatic abort sequence to begin.

The launch escape system pulled the Soyuz Descent Module clear from the booster, and it then made an emergency touchdown, landing in the snow-covered Altai Mountains. The capsule's parachute apparently caught in a tree right over a gorge, at the end of a mission lasting only 22 minutes. Upon his return, Makarov was awarded his second Order of Lenin, for 'courage displayed during the launch.'

The aborted mission was subsequently designated 'the April 5 anomaly', and the back-up crew flew the replacement mission – Soyuz 18.

Makarov made a third spaceflight as flight-engineer onboard Soyuz 27, launched on 10 January 1978. The spacecraft docked with the orbiting Salyut 6 space-station, and Makarov and Vladimir Dzhanibekov became the first crew to visit an already orbiting crew. During their 5 day stay onboard the space-station, the two cosmonauts carried out a series of joint experiments with the host crew of Gubarev and Grechko. They then returned to Earth after a flight lasting 5 days 22 hours and 59 minutes.

Makarov was subsequently awarded the title of Hero of the Soviet Union for the second time, and also received his third Order of Lenin.

His next assignment was as the back-up to Vladimir Aksyonov, the flight-engineer onboard Soyuz T-2, launched on 5 June 1980, and his next spaceflight came as flight-engineer onboard Soyuz T-3, launched on 27 November 1980. The flight was the first test flight of the three man version of the Soyuz T spacecraft. It docked with the Salyut 6 space-station, and Makarov, Gennady Strekalov and Leonid Kizim spent 10 days carrying out repairs to the station before returning to Earth after a mission lasting 12 days 19 hours and 8 minutes.

Makarov was subsequently awarded his fourth Order of Lenin.

He is married, and he and his wife Valentina, who works in a design office, have two sons, Kostya, born in 1974 or 1975, and Leonid. Makarov is a keen chess-player, hockey fan, theatre-goer, and avid reader. He is also a confirmed tourist, and enjoys travelling around the Soviet Union. In addition, he is involved in the Soviet-Ethiopia Friendship Society. In June 1978, he was presented with a Gold Medal by the Mayor of Lyons in recognition of the Soviet Union's role in the exploration of space.

Other cosmonauts have described him as uncompromising and firm, and Dzhanibekov once said of him 'Oleg Makarov is an experienced, skilled engineer. He always acts faultlessly....At first sight, Oleg looks a tough man, but I believe that this way he conceals his inborn tact.'

Said to be a brilliant engineer, Makarov has been personally involved in the designing and construction of all Soviet manned spacecraft, including the Salyut space-stations and the Soyuz T. He later said that the April 5 anomaly experience made him regard his work as an engineer even more strictly. Including the April 5 anomaly, he was the first cosmonaut to fly four missions.

Oleg Makarov was the 27th Soviet cosmonaut, and he has spent a total of 20 days 17 hours and 45 minutes in space.

Lieutenant-Colonel Ravish Malhotra

Ravish Malhotra was born on 25 December 1943, in Calcutta, India. He completed a course of studies at the St.Thomas High School in Calcutta, and then enrolled as an Air Force cadet at India's National Defence Academy, at Khadakvasla. Upon graduation, he was commissioned into the Indian Air Force, and later qualified as a test-pilot.

In June 1979, the Soviet Union offered India the chance to fly a cosmonaut onboard a joint Soviet-Indian spaceflight, along the same lines as the agreement to fly non-Soviet cosmonauts from member countries of the Intercosmos programme. After further discussions, the offer was accepted by Indian Prime Minister Gandhi in March 1981, and India began a cosmonaut selection programme. The two finalists were Malhotra and Rakesh Sharma and they were sent to begin their training at the Yuri Gagarin Cosmonaut Training Centre, arriving on 19 September 1982. At the time of his selection, Malhotra had 3400 hours flying time to his credit.

In June 1983 the two men were taken to the Black Sea port of Fedosia to practice splashdown procedures. During these exercises, photographs were released showing Malhotra training with Yuri Malyshev and Nikolai Rukavishnikov, and Sharma with Anatoli Berezovoi and Georgi Grechko. It was assumed that these were the crews which would make the actual flight to Salyut 7, with the Malhotra crew as the prime choice, as Malyshev had originally been scheduled to command the joint Soviet-French mission before he was stood down.

In early July the two cosmonauts arrived back in India for a months holiday, and upon their return to Star Town in August, were paired with their respective commanders and flight-engineers.

Finally, on 6 October 1983, the crews for the joint flight were officially named. The prime crew consisted of Malyshev, Rukavishnikov (subsequently replaced by Gennady Strekalov) and Sharma; the reserve crew was Berezovoi, Grechko and Malhotra. Therefore, for some reason Malhotra had been replaced as prime by Sharma. According to the Indian press it was because Sharma had a better command of the Russian language.

Sharma eventually flew the mission as research-cosmonaut onboard Soyuz T-11, launched on 3 April 1984, while Malhotra served as back-up, and also as CapCom in FCC.

Following the successful completion of the mission, Malhotra was awarded the Soviet Order of People's Friendship, and the Indian Kirti Chakra Order.

Ravish Malhotra is married, and he and his wife, Mira, formerly a hotel manager, have a daughter, Rakhi, born in 1972 or 1973, and a son, Rohid, born in 1974 or 1975. Malhotra's hobbies include skiing, which he learned during his two winters at Star Town.

Colonel Yuri Vasilyevich Malyshev

Yuri Malyshev was born on 27 August 1941, in the city of Nikolayevsk, near Volgograd. After finishing secondary school, he enrolled in the Kachinsk Flying School, and then went on to the Higher Air Force School in Kharkov, in the Ukraine, graduating in 1963. He then served in the Soviet Air Force, was trained as a fighter-bomber pilot, and flew MiG-17 and MiG-21 fighters. He joined the CPSU in 1964.

After Gagarin's flight, Malyshev's thoughts turned to outer space. But apparently, he didn't hurry to file an application, because at the age of only 19, he thought it wiser to gain more flying experience first. When he did apply, he was successful, and was accepted into the cosmonaut team in 1967, reporting for duty in May.

It is believed that he joined a special training group in 1976, preparing for solo Soyuz missions, and his first assignment was as back-up to Valery Bykovsky, the commander of Soyuz 22, launched on 15 September 1976. While training, he became closely acquainted with Vladimir Aksyonov, the Soyuz 22 flight-engineer, with whom he was later to make a flight.

He graduated from the Gagarin Air Force Academy in 1977, having begun his studies in 1973, and his first spaceflight came as commander of Soyuz T-2, launched on 5 June 1980. He and Vladimir Aksyonov docked with the orbiting Salyut 6 space-station and joined the crew of Popov and Ryumin already onboard. Their mission was the first manned test of the Soyuz T spacecraft, and the two men returned to Earth after a flight lasting 3 days 22 hours and 19 minutes.

Upon his return, Malyshev was awarded the title of Hero of the Soviet Union and also received the Order of Lenin.

In October 1981, he was named as commander of the joint Soviet-French spaceflight. However, in February 1982, it was revealed that he had had to step down due to 'medical reasons.' The French later disclosed that it was because of personal incompatibility. He was replaced by Vladimir Dzhanibekov, and the joint mission, Soyuz T-6, was launched on 24 June 1982.

In October 1983, he was named as commander of the joint Soviet-Indian spaceflight,

Soyuz T-11, launched on 3 April 1984. Malyshev, Gennady Strekalov and Rakesh Sharma docked with the orbiting Salyut 7 space-station and joined the crew of Kizim, Solovyov and Atkov already onboard. The three men carried out a programme of joint Soviet-Indian experiments and observations, before returning to Earth after a flight lasting 7 days 21 hours and 41 minutes.

Malyshev was subsequently awarded the title of Hero of the Soviet Union and also received the Order of Lenin, both for the second time. In addition, he received the Indian Ashoka Chakra Order.

He is married, and has two children, a daughter and a son. He is a pilot, 1st class, and test-pilot, 3rd class. His father is a war veteran, and is now a pensioner. Malyshev is Vice President of the Soviet-Nepal Friendship Society.

Yuri Malyshev was the 47th Soviet cosmonaut, and he has spent a total of 11 days 20 hours and 00 minutes in space.

Major-General Andrian Grigoryevich Nikolayev

Andrian Nikolayev was born on 5 September 1929, in the mountain village of Shorshely, in the Volga Basin, in the Soviet Union's Chuvash Autonomous Republic, some 700 miles East of Moscow, where a space museum has been built in his honour. The son of a farmer, he was one of 6 children. He first saw a plane when he was 7, and apparently talked of little else for the next few weeks, expressing his desire to fly one someday. Later however, he decided to become a doctor, and entered medical school. Under the influence of his elder brother Ivan, he switched to forestry, and transferred to a forestry school, eventually graduating as a technical forester. He joined the Soviet Air Force in 1950, training first as a gunner, then as a radio-operator. He graduated from an Air Force School in 1954, and joined the CPSU in 1957.

He was selected to join the cosmonaut team in 1960, and his first assignment was as the back-up to Gherman Titov, the pilot of Vostok 2, launched on 6 August 1961. After Titov, Nikolayev holds the record for a stay in the isolation chamber – 4 days (96 hours). While in the chamber, he painted pictures, and during special heat tests, he kept cool by painting pictures of ski-slopes and blizzards. The doctors monitoring the tests were impressed with his ability to maintain his body temperature psychologically.

For Vostok 1 and Vostok 2, the cosmonauts elected him to serve as their 'quartermaster', as they considered him 'a thoughtful, punctual and scrupulous provider of foodstuffs and drinking water.' They even elected him quartermaster for his own spaceflight !

His first spaceflight came as pilot of Vostok 3, launched on 11 August 1962, when he formed half of the world's first joint flight, consisting of Vostok 3 and Vostok 4. At one time, Nikolayev was close enough to see the Vostok 4 capsule, carrying Pavel Popovich, as they passed within 4.8 km (3 miles) of each other. Nikolayev returned to Earth after a flight lasting 3 days 22 hours and 22 minutes.

Upon his return, Nikolayev was awarded the title of Hero of the Soviet Union and also received the Order of Lenin. He then assisted with other spaceflights, and helped to develop the Soyuz spacecraft.

On 23 April 1967, Vladimir Komarov was launched onboard Soyuz 1. He was brought back to Earth after just over a day in orbit, and was killed when the spacecraft's parachute tangled following re-entry. It is now clear, despite official Soviet denials at the time, that a second spacecraft, Soyuz 2, was to have been launched on 24 April. Following docking, an exchange of crews would have taken place.

The Soyuz 2 prime and back-up crews have since been identified by Western observers. The prime crew were Valery Bykovsky, Alexei Yeliseyev and Yevgeny Khrunov, and the back-up crew were Pavel Belyayev/Andrian Nikolayev, Valery Kubasov and Viktor Gorbatko. (There is some uncertainty as to whether Nikolayev or Belyayev was the back-up Soyuz 2 commander.) Following the initial problems with Soyuz 1, the Soyuz 2 launch was cancelled.

An unmanned Soyuz 2 subsequently flew in a rendezvous operation with Soyuz 3, piloted by Georgi Beregovoi, and in January 1969, nearly two years late, the mission intended for Soyuz 1/2 was carried out by the Soyuz 4/5 spacecraft. The prime and back-up crews of Soyuz 2 were given the same assignments on Soyuz 5 – only the commanders had changed. The position of back-up commander was filled by Georgi Shonin.

Two pre-launch photographs have been released showing Komarov and the prime Soyuz 2 crew together with Nikolayev, leading some observers to suggest that he was the back-up commander, and not Belyayev.

However, if Nikolayev was the back-up, he should have been re-assigned together with Kubasov and Gorbatko to form the Soyuz 5 back-up crew. In fact, his next assignment was as the back-up commander for the Soyuz 6, 7 and 8 missions, all launched in October 1969, and his absence from the Soyuz 5 back-up crew would seem to indicate that he was NOT the Soyuz 2 back-up commander.

In 1968, Nikolayev graduated from the Zhukovsky Military Air Academy.

His second spaceflight came as commander of Soyuz 9, launched on 1 June 1970. The flight was a long-duration one to test the effects of prolonged weightlessness on Man. After making a successful re-entry and touchdown, Nikolayev and Vitaly Sevastyanov had to be carried from the descent module. Their record flight of 17 days 16 hours and 59 minutes onboard the cramped Soyuz spacecraft had left them weak, and they were unable to walk or sleep properly for the next four or five days.

Nikolayev later confided to Valery Ryumin that when he emerged from the capsule, he felt 'as if his legs had turned into a tail. It seemed as if he had no use for them anymore.' He was subsequently awarded the title of Hero of the Soviet Union and also received the Order of Lenin, both for the second time. In addition, he was later promoted to Major-General.

In July 1971, he caused a major security scare in Paris when he vanished from a cocktail party being held in the Soviet Embassy. There was immediate panic, as it was thought that he must have defected. However, he was later found enjoying a quiet drink in a nearby café He was taken back to the Embassy and then put on the first plane back to Moscow in the morning.

In the period 1971 – 73, Nikolayev is thought to have worked on Salyut development, probably on the military programme. In July 1974, he acted as a flight-controller for the Soyuz 14/Salyut 3 mission, and in December 1974, during the Soyuz 16 mission, he acted as 'deputy mission commander for the training and work of the crew', which is understood to have been a training role, possibly as a deputy to Valery Bykovsky, the ASTP Training Manager.

In January 1975, he acted as a flight-controller for the Soyuz 17/Salyut 4 mission. Also in

1975, he was awarded the degree of Candidate of Technical Sciences. Nikolayev, known to his friends as Andrei, is presently First Deputy Head of the Yuri Gagarin Cosmonaut Training Centre. He is a Deputy to the Supreme Soviet of the Russian Federation, the largest of the Republics of the Soviet Union.

Titov once described him as 'a man of iron endurance and courageous determination . . . the embodiment of composure – the calmest man in an emergency that I have ever known.' His favourite expression is apparently 'The main thing is to keep calm,' and he is said to have an unflappable nature. Once during a flight in a MiG, the engine cut out, and all attempts to re-start it proved fruitless. Ground control ordered him to abandon the aircraft, but he simply replied 'The main thing is to keep calm.' When told that he would not be able to reach the airstrip, he replied that he would land on a field. Although it seemed impossible that he would survive, he managed a safe forced landing, and walked away without a scratch, and began noting down on a pad what had gone wrong with the jet! Ever since, he has had the reputation of being an unusually self-possessed and cool-headed person – 'a model of reserve and icy composure.'

In November 1963, he married Valentina Tereshkova, the Soviet Union's first woman in space, but in June 1983, it became known that the couple had divorced. They have a daughter, Yelena, (also known as Alyonka) born on 8 June 1964.

Andrian Nikolayev was the 3rd Soviet cosmonaut, and he has spent a total of 21 days 15 hours and 21 minutes in space.

Viktor Ivanovich Patsayev (1933-1971)

Viktor Patsayev was born on 19 June 1933, in Aktyubinsk, Kazakhstan. He graduated from the Penze Industrial Institute in 1955 and then worked as a radio-research worker. Later, he worked at the Central Aerological Observatory as a design-engineer, until he was selected to join the cosmonaut team in 1969. He joined the CPSU in 1968.

His only spaceflight was as test-engineer onboard Soyuz 11, launched on 6 June 1971. *Patsayev was specially selected for the Salyut 1 programme, and it is believed that he was always intended to fly on the second mission to the space-station. However, due to various problems, it appears that he may have been forced to switch crews at some stage, and this seems to have been confirmed by Soviet reports stating that he was 'a last-minute addition to the Dobrovolsky crew in 1971.' (see section entitled 'Salyut 1 Crewing' for more details.)*

The Soyuz 11 crew docked with the Salyut 1 space-station, and spent nearly 23 days onboard, carrying out Earth observations work and other experiments. Then Patsayev, Volkov, and commander Dobrovolsky reboarded their Soyuz spacecraft and returned to Earth. Tragically, a pressure equalization valve opened prematurely while the descent module was still above the atmosphere, and the crew were killed following the loss of cabin oxygen and pressure. The death of the three men on 30 June 1971 was a severe blow to the Soviet manned space programme, and delayed further flights for over two years. The Soyuz 11 mission had lasted 23 days 18 hours and 22 minutes, a record which stood for nearly two years before being broken by the American Skylab 1 astronauts in 1973.

Patsayev was posthumously awarded the title of Hero of the Soviet Union and also received the Order of Lenin. Following a military state funeral, his ashes were buried in the Kremlin Wall. In his honour, the Soviets named one of the space tracking vessels the 'Cosmonaut Viktor Patsayev', and a street in Kaluga, the birthplace of Tsiolkovsky, was also named after him.

He was married, and he and his wife Vera Alexandrovna – a research worker – had two children, Dmitry, born in 1957, and Svetlana, born in 1962. Shortly before the Soyuz 11 flight, Patsayev had earned a Candidate of Science degree.

Viktor Patsayev was the 25th Soviet cosmonaut.

Major Oldrich Pelčak

Oldrich Pelčak (or Oldrzich Pelczak) was born on 2 November 1943, in Gottwaldov, Czechoslovakia.

He graduated from the Ugerske-Gradiste engineering junior college in 1962, and then attended the Kosice Air Force College, graduating in 1965. He joined the Communist Party of Czechoslovakia in 1964, and attended the Gagarin Air Force Academy from 1972 – 76.

Following the agreement to fly non-Soviet cosmonauts from member countries of the Intercosmos programme onboard Soyuz and Salyut spacecraft, Czechoslovakia began a cosmonaut selection programme. The two finalists were Pelčak and Vladimir Remek and they were sent to begin their training at the Yuri Gagarin Cosmonaut Training Centre in December 1976. Remek eventually flew the mission as research-cosmonaut onboard Soyuz 28, launched on 2 March 1978, while Pelčak served as back-up, and probably as CapCom in FCC.

Oldrich Pelčak is married, and he and his wife Hana have two children, Oldrich and Milos.

Colonel Leonid Ivanovich Popov

Leonid Popov was born on 31 August 1945 in the city of Alexandria, Kirovograd Region, in the Ukraine. His father Ivan Alexeiyevich Popov was a crop farmer, and chairman of a co-operative farm. He advised Popov to go to an agricultural academy after school, but instead he went to work in an electrical engineering plant for two years, where he mastered the trades of fitter and electrician.

He married in 1963, immediately after graduation from school, and he and his wife worked at the same factory.

Then one day, during a holiday, very suddenly and without a word to his family, he went to an air school to sit the entrance exams. He subsequently graduated in 1968 from the Higher Air Force College, in Chernigov, in the Ukraine, following which he served as a fighter pilot in the Soviet Air Force. He became an experienced pilot, and accumulated more than 1200 hours in the air. He had a close escape on one occasion when his plane's engine stalled, and he had to bale out.

While Popov went to the Air College, his wife went to a technical school, and after his graduation, they lived in Azerbaijan for two years.

He was selected to join the cosmonaut team in 1970, at the age of 25, and his initial training was supervised by Boris Volynov. He joined the CPSU in 1971, and in 1975, was assigned to a Salyut training group which also included Kizim and Romanenko. He graduated from the Gagarin Air Force Academy in 1976, while still undergoing training at the Yuri Gagarin Cosmonaut Training Centre.

His first assignment was as one of the back-ups to Valery Bykovsky, the commander of Soyuz 22, launched on 15 September 1976. Yuri Malyshev has been named as the main back-up, so it appears that Popov was the commander of the support crew.

It is believed that his next assignment was as commander of the support crew for what was intended to be the second long-duration flight to Salyut 6. He should then have served as back-up for the next mission. However, following the failure of Soyuz 25, the crewing schedules were revised, and it is believed that he served as commander of the support crew for Soyuz 29, launched on 15 June 1978. (see section entitled 'Salyut 6 Crewing' for more details.)

His next assignment was as the back-up to Vladimir Lyakhov, the commander of Soyuz 32, launched on 25 February 1979, and his first spaceflight came as commander of the Soyuz 35 mission, launched on 9 April 1980. He had originally trained with Valentin Lebedev for the mission, but following an injury, Lebedev was replaced by Valery Ryumin. The two men docked their spacecraft with the orbiting Salyut 6 space-station and carried out a long programme of research and experiments as well as playing host to the visiting cosmonauts from Hungary, Vietnam and Cuba.

Ryumin and Popov returned to Earth on 11 October 1980, after a flight lasting 184 days 20 hours and 12 minutes – thus setting a new duration record. Upon his return, Popov was awarded the title of Hero of the Soviet Union and also received the Order of Lenin. He was also awarded the title of Hero of the Republic of Cuba, with Gold Star Medal and the Order of Playa-Giron, together with Vietnam's highest honours. During the mission, he had celebrated his 35th birthday.

In January 1981, Popov was teamed up with Dumitru Prunariu to train for the joint Soviet-Romanian spaceflight, and the two men were launched onboard Soyuz 40 on 14 May 1981. They linked up with the orbiting Salyut 6 space-station, and joined the crew of Kovalyonok and Savinykh already onboard. They then carried out a joint programme of Soviet-Romanian experiments and observations before returning to Earth after a mission lasting 7 days 20 hours and 41 minutes.

Upon his return, Popov was awarded the title of Hero of the Soviet Union and also received the Order of Lenin, both for the second time. In addition, he was awarded the title of Hero of the Socialist Republic of Romania.

According to unofficial information, Dumitru Dediu, the back-up on Soyuz 40, was in fact, chosen by the Soviet Union to be the first Romanian in space. He was named as such in a United Press report issued in Moscow shortly before the mission began. It therefore seems that he and Valery Bykovsky should have flown, but when Bykovsky became disqualified, the back-up crew of Popov and Prunariu took the prime slot. (see section entitled 'Intercosmos Crewing' for more details.)

Popov's next spaceflight came as commander of Soyuz T-7, launched on 19 August 1982. The spacecraft docked with the Salyut 7 space-station, and Popov, Alexander Serebrov and Svetlana Savitskaya joined the crew of Berezovoi and Lebedev already onboard. The visiting crew spent nearly seven days carrying out a programme of experiments before returning to Earth after a flight lasting 7 days 21 hours and 52 minutes.

Upon his return, Popov was awarded his third Order of Lenin.

He is married, and he and his wife Valentina have two children, Yelena (Lena), the eldest, and Alexei (Alyosha), born in 1970 or 1971.

His wife once described the difficulties of being married to a cosmonaut. 'Before, I used to wait for Leonid to come back from flights. In Star Town I waited for him to come back from training sessions. That's how I lived until his first spaceflight. I tried to convince myself that it was just another long assignment, but I couldn't. On top of that Leonid got the longest flight – 185 days! Not a single one of the 185 days was an ordinary day in my life. I felt as though I had been set adrift. I'd go out into the street and see people talking and laughing, but I couldn't stop wondering how Leonid was doing out there.'

Popov is known as Lenya to his friends. When asked to describe him, Valery Ryumin spoke of his 'high professionalism – He does his job very well indeed.' Svetlana Savitskaya said he was 'A kindly, sociable person.' His hobbies include camping and fishing, and he is Vice-President of the Soviet – Laos Friendship Society.

Leonid Popov was the 46th Soviet cosmonaut, and he has spent a total of 200 days 14 hours and 45 minutes in space.

Major-General Pavel Romanovich Popovich

Pavel Popovich was born on 5 October 1930, in Uzin, near Kiev. After finishing secondary school, he went to a trade school, and then to an industrial secondary school at Magnitogorsk, graduating in 1951 as a carpenter. He had become very interested in flying, and after graduating with distinction from the local aeroclub, he went to an Air Force school, graduating in 1954. He then served in the Soviet Air Force as a pilot for a number of years, and was awarded the Order of the Red Star for an undisclosed 'government assignment' in the Soviet Arctic. He joined the CPSU in 1957.

On 11 January 1960, he was the very first member of the new cosmonaut team to report for duty, and greeted Nikolayev, Titov and Gagarin when they arrived, helping them to settle in. He was very popular with his fellow cosmonauts, who, it is said, all strongly admired his fine tenor voice, which he put to good use in his constant singing. One story told how he had spent over two weeks in the isolation chamber during training, and that ' he sang from morning till night, whistled his favourite Ukrainian tunes, and even danced. He was a restless type, with a lively, dynamic personality. One sometimes had the impression that nothing could make Pavel Popovich sit still for as much as an hour or two.'

In 1960, he was elected as leader of the CPSU in Star Town for the cosmonaut team, and in April 1961, he served as CapCom for Vostok 1.

His first spaceflight came as pilot of Vostok 4, launched on 12 August 1962, when he formed half of the world's first joint flight, consisting of Vostok 3 and Vostok 4. He returned to Earth after a flight lasting 2 days 22 hours and 57 minutes.

Upon his return, Popovich was awarded the title of Hero of the Soviet Union and also received the Order of Lenin.

In mid-1967, he was assigned to the Zond training group, and began preparing for a flight to the Moon. He graduated from the Zhukovsky Military Air Academy in 1968, and upon the abandonment of the Soviet lunar programme, he was re-assigned to the Soyuz programme, serving as CapCom for the joint Soyuz 6/7/8 mission in October 1969.

In March 1972, a Western news agency reported that two Soyuz missions were to be launched to a new Salyut in May of that year, and Popovich was named as a back-up crewmember. The report is a strange one, particularly as it also named two rookie cosmonauts – Vinogradov and Voronov – who have never flown.

It is also thought that Popovich may have trained for Salyut 2, but when that failed, he was re-assigned to Salyut 3. (see section entitled 'Salyut 2 / Cosmos 557 Crewing' for more details.)

He had to wait 12 years before making a second spaceflight, as commander of Soyuz 14, launched on 3 July 1974. He and Yuri Artyukhin had begun joint training for the mission in August 1973, and following launch, they docked with the orbiting Salyut 3 space-station and spent nearly two weeks onboard. They carried out a programme of Earth observations and bio-medical experiments before returning to Earth, after a flight lasting 15 days 17 hours and 30 minutes.

Popovich was subsequently awarded the title of Hero of the Soviet Union and also received the Order of Lenin, both for the second time.

In July 1976, he served as CapCom for the Soyuz 21 / Salyut 5 mission, and in 1978, he was awarded a Candidate of Technical Sciences degree. He is presently Deputy Head of the Yuri Gagarin Cosmonaut Training Centre.

In February 1984, he was appointed to head a Commission set up to study abnormal atmospheric phenomena – i.e. UFO's.

He is actively involved in public affairs, and is a deputy to the Supreme Soviet of the Ukraine, a member of the Union of Writers of the Ukraine, Vice-Chairman of the Soviet – Austria Friendship Society, and Chairman of the Soviet Boxing Federation. He has written an autobiographical book entitled 'Take-off in the Morning', and also another book entitled 'Test By Earth & Space' published in 1982. In addition to his reputation for cheerfulness, he is said to be a hard working and persistent man.

He is married, and his wife Marina Lavrentevna Vasilyeva-Popovich, born in 1931, was, at one time, a Colonel in the Soviet Air Force, and a famous test-pilot who had set numerous world records. At the time of Popovich's first spaceflight, she had logged more flying hours than him ! They first met in 1951, at an evening party for the flying school students, and took to one another at once. However, they didn't meet again until four years later, when they married. They then moved North, where Popovich had been posted. They have two daughters, Natasha and Oksana.

Marina Popovich is nearly as well-known in the Soviet Union as her husband. A Candidate of Science (Engineering), she had in her career as a test-pilot been connected mainly with Oleg Antonov's planes, having tested many of his prototypes, from the smallest to the largest.

In 1964, flying a sports jet, she set her first world airspeed record, flying around a 100 km triangular course. On 23 February 1972, flying the AN22 Antaeus, she and her crew broke ten world records in two flights covering 2000 km with loads of 30, 35, 45 and 50 tons.

As a student in Novosibirsk, she joined the parachute section of an amateur flying club. She wanted to become a pilot, and for the next 3 or 4 years she virtually lived on the doorsteps of the Air Chiefs – getting a flat refusal every time. So she went to Moscow, gate-crashed the offices of the Supreme Soviet, and secured an interview with President Voroshilov.

He was evidently impressed, for when she told him how much she wanted to become a pilot, he raised no objection, and there and then wrote to Air Force Lieutenant-General Kamanin, Chairman of the Army, Navy and Air Force Voluntary Aid Commitee. The note simply said ' Please consider this application. If she has the makings, send her on a course.' She subsequently proved her point when some years later she was awarded the Gold Medal of the International Aeronautical Federation (IAF).

Pavel Popovich was the 4th Soviet cosmonaut, and he has spent a total of 18 days 16 hours and 27 minutes in space.

Major-Engineer Dumitru Dorin Prunariu

Dumitru Prunariu was born on 27 September 1952 in the city of Brasov, Romania, into the family of an automobile engineer. He was apparently greatly interested in the flight of Sputnik 1, and was awarded the Gold Medal at an international exhibition of children's technical drawings for the design of a launching device. He finished an eight-year secondary school and went on to the school of Physics and Mathematics in Brasov before enrolling in the aircraft engineering department at the Bucharest Polytechnical Institute. He joined the Romanian Communist Party in 1973.

Upon graduation from the Institute in 1976, he worked as an engineer at the Brasov aircraft factory. He then enrolled in the Air Force Officers Training School, and graduated in 1977 with the rank of Junior-Lieutenant. He became a pilot, first-class, and served in the Romanian Air Force before being selected for spaceflight training.

Following the agreement to fly non-Soviet cosmonauts from member countries of the Intercosmos programme onboard Soyuz and Salyut spacecraft, Romania began a cosmonaut selection programme. The two finalists were Prunariu and Dumitru Dediu, who began their training at the Yuri Gagarin Cosmonaut Training Centre in March 1978.

In January 1981, Prunariu was teamed up with Leonid Popov in preparation for the joint flight. He eventually flew as research-cosmonaut onboard Soyuz 40, launched on 14 May 1981. The two men docked with the orbiting Salyut 6 space-station, and joined the crew of Kovalyonok and Savinykh already onboard. They then carried out a joint programme of Soviet-Romanian experiments and observations before returning to Earth after a mission lasting 7 days 20 hours and 41 minutes.

Upon his return, Prunariu was awarded the title of Hero of the Soviet Union and also received the Order of Lenin. In addition, he was awarded the title of Hero of the Socialist Republic of Romania.

According to unofficial information, Dumitru Dediu, the back-up on Soyuz 40, was in fact, chosen by the Soviet Union to be the first Romanian in space. He was named as such in a United Press report issued in Moscow shortly before the mission began. It therefore seems that Valery Bykovsky and Dediu should have flown, but when Bykovsky became disqualified, the back-up crew of Popov and Prunariu took the prime slot. Dediu was moved to back-up, and Yuri Romanenko became the new back-up commander. This switch of crews was also reported by United Press.

Prunariu is married, and he and his wife, Ursu Rodiche-Crine, an aircraft designer, have two sons, born in 1975 or 1976, and in 1977 or 1978. His favourite hobby is photography. Alexei Leonov once described him as 'a serious, well-trained man liked by everybody . . . an extraordinary man who coped with all difficulties, particularly during the training.'

Dumitru Prunariu was the 9th Intercosmos cosmonaut.

Colonel-Engineer Vladimir Remek

Vladimir Remek was born on 26 September 1948 in Ceske-Budejovice, South Bohemia, Czechoslovakia. He attended nine-year elementary schools in Ceske-Budejovice and Brno, and then went on to a secondary school at Caslav in 1963, where he specialised in maths and physics. While at school, he became involved in the youth movement, and took up aircraft-modelling as a hobby. He was also very keen on reading and sports, and excelled at running.

In 1966, he graduated from Caslav school, and turned down the prospect of a career in nuclear physics to join the Air Force – his burning ambition. He enrolled at the Air Force College at Kosice, East Slovakia, where he received training in supersonic fighter flying, and became a career officer. He did his flight training between 1966 – 70 onboard L-29 test jets, and later transferred to supersonic aircraft.

During his second year at the Air Force college, he was accepted as a member of the Communist Party of Czechoslovakia, thanks to 'good results at work, excellent study work, and firm political viewpoints.' He graduated from the college in 1970 with the rank of Lieutenant.

Following graduation, he was assigned to the Zvolensky combat unit, where he advanced, in a very short time, to the grade of Pilot – second class. As the best pilot in his squadron, he was offered a scholarship at the Gagarin Air Force Academy in the Soviet Union, and commenced his studies on 1 September 1972.

In 1975, he was awarded the Medal for Service to the Motherland, and in 1976 he graduated from the Gagarin Air Force Academy. He returned to his unit in Czechoslovakia as a First-Lieutenant, and was soon promoted to Captain (in November 1976.)

He returned to Czechoslovakia very shortly after the agreement to fly non-Soviet cosmonauts from member countries of the Intercosmos programme onboard Soyuz and Salyut spacecraft was announced. Czechoslovakia began a cosmonaut selection programme, and he was one of the dozens of pilots picked in July 1976 as candidates for the programme. After months of medical examinations and tests, the men were short listed and a group were sent to Star Town in November 1976. After two weeks of tests, the two finalists were chosen – Remek and Oldrich Pelčak – and they were sent to begin their training at the Yuri Gagarin Cosmonaut Training Centre in December 1976.

During his training period, he would often sleep with his bed propped up by a pile of books, to keep his head lower than his feet, to help prepare him for the increased blood flow to the head experienced in weightlessness. During his first training session in a Soyuz simulator, Remek lost 2kg (4.4 lbs) in weight, during a three-hour session wearing a spacesuit.

He was paired with Alexei Gubarev in May 1977, and the two men began joint training the following August. At the end of February 1978, Remek and Pelčak, together with the Polish and GDR cosmonauts, sat their final exams and tests at the Yuri Gagarin Cosmonaut Training Centre. The results were extremely close, but eventually, Remek was selected to be the first representative of the Intercosmos countries to fly in space.

Only one month later, on 2 March 1978, he was launched as research-cosmonaut onboard Soyuz 28. He and Alexei Gubarev docked with the orbiting Salyut 6 space-station, and joined the crew of Romanenko and Grechko already onboard. They then spent nearly seven days carrying out a series of joint Soviet-Czech experiments and observations, before returning to Earth after a flight lasting 7 days, 22 hours and 16 minutes.

Upon his return, Remek was awarded the title of Hero of the Soviet Union and also received the Order of Lenin. He was also awarded the titles of Hero of the Czechoslovak Socialist Republic and Cosmonaut-Flier of the Czechoslovak Socialist Republic. He has since written a book about the flight entitled 'Planet Earth Below Us', and has also co-authored another book with Alexei Gubarev entitled 'International Orbit', published in 1983.

Remek married in early 1979, and his wife, Hana Davidova, was a Prague actress. They subsequently divorced. Remek's father, Josef, is a retired Lieutenant-General in the Czech Air Force, and was believed to be a deputy Defence minister in the Czech government. His mother is Mrs. Blanka Remekova, and he has two sisters.

Vladimir Remek was the 1st Intercosmos cosmonaut.

Colonel Yuri Victorovich Romanenko

Yuri Romanenko was born on 1 August 1944, in the settlement of Koltubanovsky, Buzuluksky district, Orenburg Region, in the Middle Volga area. His father was a ship's captain, and so Romanenko's childhood was spent on the shores of the northern Russian seas. At school, he displayed a keen interest in mathematics, physics, model ships, and later, in model aircraft. On finishing secondary school, he worked for a year as a fitter, before enrolling in the the Higher Air Force College, in Chernigov, in the Ukraine in 1962. He graduated with honours in 1966, with the diploma of engineer-pilot.

He joined the CPSU in 1965, and from 1966 onwards he served as a flight-instructor in the Soviet Air Force at the Higher Air Force College, in Chernigov. He logged over 700 hours flying time and became an ace pilot. A visit by cosmonaut Gherman Titov to his regiment is said to have convinced Romanenko that space would be the next step in his career.

He married in 1969, when he was a Lieutenant in the Soviet Air Force, and his wife was 18 when they married. Less than two months later, Romanenko was summoned to Moscow, and was chosen for cosmonaut training. That was in 1969, although his official selection date is 1970.

His first assignment came in May 1973, when he was named as back-up to Anatoly Filipchenko, the commander of the reserve Soyuz for the Soviet half of the Apollo-Soyuz Test Project. From May 1973 he trained for the mission in both the United States and the Soviet Union. As part of the preparations for the joint flight, Romanenko served as back-up to Filipchenko on the Soyuz 16 ASTP rehearsal mission, launched on 2 December 1974. In July 1975, he served as one of the Soviet Capcoms during the Apollo-Soyuz mission.

Following the completion of the flight, Romanenko was assigned to a Salyut training group which also included Kizim and Popov.

His next assignment was as back-up to Vladimir Kovalyonok, the commander of Soyuz 25, launched on 9 October 1977. Following the failure of that mission, Romanenko was re-assigned as commander of Soyuz 26, the replacement mission. He and Georgi Grechko were launched onboard Soyuz 26 on 10 December 1977, and docked with the Salyut 6 space-station. The prime objective of the mission was to examine the station's docking mechanism, as it was felt that it might have been damaged during Soyuz 25's abortive docking attempts. Soyuz 26 docked at the station's rear docking port, and Romanenko and Grechko carried out an EVA to thoroughly examine the suspect forward docking port. They could find no damage.

Romanenko was supposed to stay in the airlock while Grechko performed the EVA. However, he wanted to see the view from outside and performed an unauthorized exit of the space-station. Unfortunately, he lost his grip and according to one report, floated off. To his horror, he realized that his safety line had not been attached. Grechko, in his own words, was able to grab the line 'near the end' and haul Romanenko back in!

During the EVA, Grechko spent 20 minutes outside Salyut, and both men were in 'the near vacuum conditions of space' for 1 hour and 28 minutes, which is understood to mean hatch-opening to hatch-closing. After re-entering the Salyut, the two settled down to their long programme of scientific experiments and observations before returning to Earth after a mission lasting 96 days and 10 hours precisely.

Upon his return, Romanenko was awarded the title of Hero of the Soviet Union and also received the Order of Lenin. In addition, he was awarded the title of Hero of the Czechoslovak Socialist Republic.

He was then placed in charge of one of the ground flight support groups for Soyuz 29, launched on 15 June 1978.

His next assignment was as back-up to Nikolai Rukavishnikov, the commander of the joint Soviet-Bulgarian mission, Soyuz 33, launched on 10 April 1979. For his part in the preparations for the mission, Romanenko was awarded the Order of the People's Republic of Bulgaria, First Class.

In September 1979, he teamed up with the Cuban, Arnaldo Tamayo-Méndez in preparation for the forthcoming joint Soviet-Cuban mission, and his second spaceflight came as commander of Soyuz 38, launched on 18 September 1980. He and Tamayo-Méndez docked with the orbiting Salyut 6 space-station, and joined the crew of Popov and Ryumin already onboard. The two men carried out a joint programme of Soviet-Cuban experiments and observations before returning to Earth after a mission lasting 7 days 20 hours and 43 minutes.

Upon his return, Romanenko was awarded the title of Hero of the Soviet Union and also received the Order of Lenin, both for the second time. In addition, he was awarded the title of Hero of the Republic of Cuba, with Gold Star Medal and Order of Playa Giron.

His next assignment was as back-up to Leonid Popov, the commander of the joint Soviet-Romanian mission, Soyuz 40, launched on 14 May 1981. *It appears that Romanenko was not the original choice for the back-up commander. According to unofficial information, Dumitru Dediu, the back-up research-cosmonaut on Soyuz 40, was in fact, chosen by the Soviet Union to be the first Romanian in space. He was named as such in a United Press report issued in Moscow shortly before the mission began. It therefore seems that Valery Bykovsky and Dediu should have flown, but when Bykovsky became disqualified, the back-up crew of Popov and Prunariu took the prime slot. Dediu was moved to back-up, and Yuri Romanenko became the new back-up commander. This switch of crews was also reported by United Press.*

Romanenko's next assignment was as the original back-up to Leonid Popov, the commander of Soyuz T-7, launched on 19 August 1982. However it appears that

Romanenko became physically disqualified shortly before the flight, and was replaced by Vladimir Vasyutin.

Romanenko and his wife Alevtina Ivanovna, a high school music teacher, have two sons – Toman (or Roman) born in 1971, and Artem, born in 1977. When his wife was asked for her reactions to her husband's occupation, she said 'It is hard to be the wife of a man who is so busy. There's all the housework to do, and the children to take care of besides my own work. All the wives in Star Town have jobs.'

'He started to train for his 96 day flight. Even on Earth, we had never been away from each other for so long. And there I was all alone with two children. Our younger son, Artem was just a few months old.'

'I took the second flight a little easier. It only lasted a week, and after that I knew exactly what being a cosmonaut's wife was like.'

'I never tried to talk my husband out of being a cosmonaut, but I don't want the children to follow in his footsteps. Of course, Yuri wants to see them become fliers. He has already infected them with his love for the sky. Right now, he is interested in underwater diving, and so are they.'

Romanenko enjoys underwater spearfishing, constructing powered model aircraft, drawing, painting and driving, and he has been described as having a 'volatile' personality. As well as the awards and honours associated with his spaceflights, he holds the following:- the 20th Victory Anniversary in the Great Patriotic War, and the 50th Anniversary of the Soviet Armed Forces, for irreproachable service, III degree.

Known to his friends as Yura, he graduated from the corresponding department of the Gagarin Air Force Academy in 1980, at which time he was said to be the youngest Colonel in the cosmonaut team. He participates actively in social affairs, and is a member of the Central Committee of the Leninist Young Communist League, and is Vice-President of the Soviet-Madagascar Friendship Society.

Yuri Romanenko was the 42nd Soviet cosmonaut, and he has spent a total of 104 days 06 hours and 43 minutes in space.

Colonel-Engineer Valery Ilyich Rozhdestvensky

Valery Rozhdestvensky was born on 13 February 1939, in Leningrad. In 1956, he enrolled at the F.E. Dzerzhinsky Higher Naval Engineering School in Leningrad, and graduated in 1961 with a degree in naval engineering. He subsequently served in the Soviet Navy, and was appointed commander of a deep-sea diving group which formed part of the rescue service of the Baltic Fleet, and at one time was an instructor in paratrooper training. He joined the CPSU in 1961.

Rozhdestvensky was selected to join the cosmonaut team in October/November 1965. *He has been linked with Salyut 3, and it is believed that his first assignment was as the flight-engineer on the support crew for Soyuz 14 and Soyuz 15, launched on 3 July and 26 August 1974, respectively. (see section entitled 'Salyut 3 Crewing' for more details.)*

His next assignment was as the back-up to Vitaly Zholobov, the flight-engineer onboard Soyuz 21, launched on 6 July 1976, and his first spaceflight came as flight-engineer onboard Soyuz 23, launched on 14 October 1976. The mission called for a docking with the Salyut 5 space-station, but this was aborted when an approach system failure occurred. Rozhdestvensky and Vyacheslav Zudov were forced to make an emergency return to Earth, splashing down in Lake Tengiz in the middle of a snowstorm, after a flight lasting only 2 days and 7 minutes.

Ironically, in a pre-flight interview, Rozhdestvensky had said that he thought that his deep-sea diving experience might stand him in good stead during the flight! Commenting on the splashdown afterwards, Zudov said it was entirely appropriate, since out of the entire cosmonaut team, Rozhdestvensky was 'the one seaman.' The two men had been formed as a crew two years earlier.

Upon his return, Rozhdestvensky was awarded the title of Hero of the Soviet Union and also received the Order of Lenin. He subsequently served as CapCom for the Soyuz 26, 29, 31, 32, 35, 36, 38, 39 and 40 missions, and is also known to have worked in FCC during the Soyuz T-9 mission.

He is married, and he and his wife, Svetlana, an engineer at Star Town, have a daughter, Tanya. He is Vice-President of the Soviet-Angola Friendship Society.

Valery Rozhdestvensky was the 38th Soviet cosmonaut

Nikolai Nikolaievich Rukavishnikov

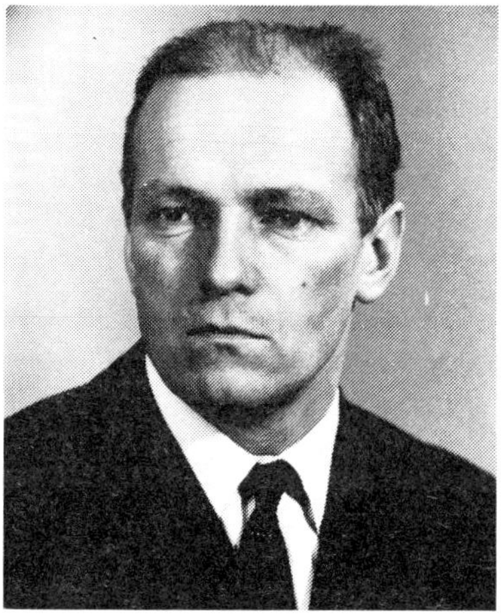

Nikolai Rukavishnikov was born on 18 September 1932, in the city of Tomsk, Western Siberia. During his childhood and adolescence, he travelled all over the Soviet Union with his parents, as his stepfather and his mother Galina, were both railroad surveyors. In doing so, he became accustomed to the hardships of camp life. In 1951, upon finishing secondary school, he enrolled at the Moscow Institute of Physics and Engineering, where he studied automation and instrument making. Upon graduating in 1957, he went to work in the design bureau headed by Sergei Korolyov. A 'creatively-minded engineer and competent designer', he helped develop automatic control systems, but did not find the work sufficiently challenging. Instead, he wanted to test the instruments and systems in space himself, so he applied to join the cosmonaut team.

He was selected to join the cosmonaut team on 23 May 1966, and arrived in Star Town in November of that year, although he did not start training until 8 January 1967. He became a leading specialist in experiments related to the Earth and Solar Physics, and in mid-1967, was assigned to the Zond training group, and began preparing for a flight to the Moon.

According to one report, he was assigned as flight-engineer for the third lunar landing mission (see section entitled 'The Zond programme' for more details.)

He joined the CPSU in 1970, and upon the abandonment of the Soviet lunar programme, he was re-assigned to the Soyuz programme. His first assignment came as the test-engineer onboard Soyuz 10, launched on 23 April 1971. *It is possible that Rukavishnikov may originally have trained for the mission with Alexei Leonov and Valery Kubasov, but they were stood down when Kubasov became ill. They were replaced by Shatalov and Yeliseyev, and Rukavishnikov bumped their test-engineer. This seems to have been confirmed by photographs showing Rukavishnikov training with Leonov and Kubasov, and Soviet reports stating that he was 'a late addition to the Soyuz 10 crew.' (See section entitled 'Salyut 1 Crewing' for more details.)*

The Soyuz 10 spacecraft docked with the Salyut 1 space-station – the first time such an operation had taken place – but undocked after only 5 hours and 30 minutes, without the crew entering the station. The Soviets claimed that it was never their intention to man the Salyut during this initial step, yet Rukavishnikov was described as 'being responsible for the operation of the Salyut systems.'

During the flight, he had reported feeling ill due to the increased flow of blood to the head experienced in weightlessness, and it was suggested at one time that his ill-health may have been the reason for cutting short the flight. Rukavishnikov has denied this however, and it would therefore seem likely that there was some problem with the transfer hatch, preventing the crew from entering the Salyut. Rukavishnikov, Vladimir Shatalov and Alexei Yeliseyev returned to Earth after a flight lasting 1 day 23 hours and 46 minutes.

Upon his return, Rukavishnikov was awarded the title of Hero of the Soviet Union and also received the Order of Lenin. *It is thought that he may have served as a back-up crewmember for the Cosmos 557 Salyut attempt in May 1973. (See section entitled 'Salyut 2/ Cosmos 557 Crewing' for more details.)* His next assignment also came in 1973, when he was named as the flight-engineer for the reserve Soyuz for the Soviet half of the Apollo-Soyuz Test Project. From May 1973 he trained for the mission in both the United States and the Soviet Union. As part of the preparations for the joint flight, Rukavishnikov made a second spaceflight as flight-engineer onboard the Soyuz 16 ASTP rehearsal mission, launched on 2 December 1974. He and Anatoly Filipchenko carried out a dress rehearsal of the ASTP Soyuz flight programme, including important docking tests, before returning to Earth after a flight lasting 5 days 22 hours and 24 minutes.

Rukavishnikov was subsequently awarded the title of Hero of the Soviet Union and also received the Order of Lenin, both for the second time.

In September 1976, he was closely involved in the Soyuz 22 mission.

His next assignment was as back-up to Alexei Gubarev, the commander of the joint Soviet-Czechoslovakian spaceflight Soyuz 28, launched on 2 March 1978.

He made his third spaceflight as commander of the joint Soviet-Bulgarian spaceflight Soyuz 33, launched on 10 April 1979. He and Georgi Ivanov were scheduled to dock with the Salyut 6 space-station and spend a week visiting the crew already onboard. However, the two men were forced to return to Earth prematurely when the Soyuz approach engine malfunctioned during the final stages of the rendezvous manoeuvres. Following an emergency ballistic re-entry, they returned to Earth after a flight lasting only 1 day 23 hours and 1 minute.

It wasn't until October 1983 that the true extent of the failure and the peril that the crew had been in, was revealed. The failure occurred in the main 917lb thrust rocket engine used for both rendezvous manoeuvres and the re-entry burn. After it had been fired six times during rendezvous manouevres, its seventh and last planned firing shut down after only 3 seconds of a planned 6 second burn.

Ground control studied the problem, and decided that the engine should be fired again in an attempt to complete the rendezvous with Salyut 6. However, when it was re-started it shut down immediately, and the crew reported feeling unusual vibrations during the firing sequence. At the time, the Soyuz was only 3000 feet (914 m) from the space-station, and the Salyut cosmonauts reported that they had seen the engine fire, but with 'a lateral glow in the power compartment.'

This report caused great alarm at ground control, as it was feared that the abnormal plume could have caused serious internal damage to the 904lb thrust back-up engine located next to the main engine. If the back-up engine failed to work, it would mean that no re-entry burn was possible, and the crew would be marooned in orbit. Calculations showed that Soyuz 33's orbit would decay naturally after 10 days, but the crew had less than 5 days of oxygen remaining.

After a worrying night considering all the possibilities, the decision was taken to attempt a re-entry burn using the back-up engine. During the 188 second burn, the engine malfunctioned as expected, but instead of shutting down early, it kept burning beyond the planned cut-off point, and started firing at less than full thrust. Eventually, after 25 seconds of excessive burn time, the crew shut down the engine manually.

The overburn was serious, as it could have resulted in a dangerously steep re-entry angle, during which the descent module might burn up. In the event, Soyuz 33 flew a highly stressed ballistic re-entry in which the cosmonauts experienced 8-10g loads compared with the 3-4g's experienced on normal re-entries. As a result of the overburn, the descent module missed the landing window, falling short of the touchdown site by several hundred miles.

Rukavishnikov was subsequently awarded the Order of Lenin for the third time, together with the title of Hero of the Peoples Republic of Bulgaria, with Gold Star Medal and the Order of Georgi Dimitrov. He was also presented with the Honorary Badge of Sofia, 1st Class, and the Badge of Honour of the town of Botevgrad, Bulgaria. In addition, he was made an honorary citizen of Omourtag and Lovech, Bulgaria.

His next assignment came in October 1983 when he was named as the flight-engineer for the joint Soviet-Indian spaceflight (Soyuz T-11, launched on 3 April 1984). The mission would have been his fourth spaceflight, but in March 1984 it was announced that he had been taken ill, and had been replaced by Gennady Strekalov. No details of his illness were given. The flight would have been his third attempt to board a Salyut, the previous two having failed.

Rukavishnikov is married, and his wife Nina Vasilyevna, a housewife, was formerly a senior technician at an engineering plant. They have a son Vladimir, born in 1965. Rukavishnikov is an amateur radio operator, and also enjoys driving. In 1983, he was elected Chairman of the Federation of Cosmonauts. He has received many awards and honours apart from those already listed, and these include the Order of Suche-Bator; the Gold Star of Hero of the People's Republic of Mongolia; the Gold Medal of the Ukrainian Academy of Sciences; the Tsiolkovsky Gold Medal of the Soviet Academy of Sciences; the Minor Gold Medal of ENEA; and the Medal 'for valiant Labour in honour of Lenin's centenary'. He is also entered in the Book of the ICLCC, and is an honorary citizen of Arkalyk, Kaluga, Karaganda, Kayhtiy, Suche-Bator and Tomsk. He also holds a degree as a Candidate of Technical Sciences.

Nikolai Rukavishnikov was the 23rd Soviet cosmonaut, and he has spent a total of 9 days 21 hours and 11 minutes in space.

Valery Viktorovich Ryumin

Valery Ryumin was born on 16 August 1939, in Komsomolsk-on-Amur, in Siberia. After graduating from a technical college in Kaliningrad, he served in the Soviet Army. In 1961, upon completing his military service, he entered the Electronics and Computer Technology Department at the Moscow Forestry Institute.

Upon graduation, he went to work in Sergei Korolyov's design bureau, where he participated in the development and creation of new space equipment, and took part in controlling the flights of manned spacecraft and orbital stations. He apparently distinguished himself as an engineer with great knowledge and initiative, and became an expert on electronic control systems.

He became one of the leading designers in the Salyut programme, and when the first Salyut station had been developed, Ryumin was made head engineer. During each Salyut launch he represented the design bureau at the space centre and was a deputy to the chief engineer for the station. He was also involved in the design of the Salyut 6 space-station.

Ryumin joined the CPSU in 1972, and was selected to join the cosmonaut team in 1973. His selection was said to be quite unusual as he is the only cosmonaut chosen who is a graduate of a Forestry-Engineering Institute, and he is also the only one to have served with land based troops – he was a tankman. He took part in the ground control of manned and unmanned spaceflights, and served as a CapCom onboard the Academician Sergei Korolyov, stationed off the coast of Cuba, during the Soyuz 17 and 18 missions in 1975. He was also said to have served as a shift deputy flight controller during ASTP.

His first spaceflight came as flight-engineer onboard Soyuz 25, launched on 9 October 1977. When he and Vladimir Kovalyonok attempted to dock their spacecraft with the Salyut 6 space-station, they were prevented from doing so by a fault in the docking system. They were forced to make an emergency return to Earth after a mission lasting 2 days and 45 minutes.

Ryumin was later awarded the Order of Lenin only – the first time that a cosmonaut had not been awarded the title of Hero of the Soviet Union following a first spaceflight, indicating official dissatisfaction with the failure of the mission.

Some years later, his wife revealed the effect the failure of the mission had on Ryumin: 'I saw the state he was in when he walked down the gangway from the plane. Clearly, he couldn't forgive himself for not having fulfilled the assignment. The period after that flight was the hardest in both his life and mine. Our son Vadim also fell ill at that time. I asked to be transferred to the night shift so that I could go to the hospital during the day. I used to come home well after midnight. Valery usually would sleep at this time, but at that period I always found him in the armchair, and I realised that he was going over and over that miserable docking operation, trying to find what went wrong, and not being able to.

I think that Valery felt he was caught up in a blind alley. Kovalyonok, with whom he went on that flight, was also really put out, but he is an emotional kind of guy and can find another channel to vent his feelings. But Valery is a reticent man, always deep in his own thoughts. That was why I was eager that he go on another flight.'

Following the failure of Soyuz 25, Kovalyonok and Ryumin were split up, and given new assignments. *It is believed that Ryumin served as flight-engineer on the support crew for Soyuz 26, launched on 10 December 1977. (see section entitled 'Salyut 6 Crewing' for more details.)* His next assignment was as back-up to Alexander Ivanchenkov, the flight-engineer onboard Soyuz 29, launched on 15 June 1978.

He made a second spaceflight as flight-engineer onboard Soyuz 32, launched on 25 February 1979. He and Vladimir Lyakhov docked their spacecraft with the orbiting Salyut 6 space-station and carried out a long programme of research and experiments. They were also intended to act as hosts to the visiting joint Soviet-Bulgarian flight, Soyuz 33, before its engine malfunctioned, causing the mission to be aborted.

When a radio-telescope attached to the Salyut became entangled with other equipment, Ryumin and Lyakhov had to carry out an EVA lasting 1 hour and 23 minutes, during which they successfully cut the telescope free. The two men returned to Earth after a flight lasting 175 days and 36 minutes.

Upon his return, Ryumin was awarded the title of Hero of the Soviet Union and also received his second Order of Lenin. He had celebrated his 40th birthday during the mission.

His third spaceflight came as flight-engineer onboard the Soyuz 35 mission, launched on 9 April 1980. Valentin Lebedev had originally trained for the mission, but following an injury, he lost his place on the mission, and had to be replaced. At the time of the accident, Ryumin had been instructing the crew in preparation for their flight.

Reports on why Ryumin was given the assignment so soon after returning from his 175-day mission vary, but he said it was because '...the launching day was drawing closer, and a replacement had to be picked. I was asked if I'd do it.'

Other reports said that many of the engineers who had already worked on Salyut 6 offered their services, and Ryumin was chosen. In only three weeks of joint training with Popov, he demonstrated that he was in excellent form, and instructors at the training centre remarked on 'his wonderful ability to co-ordinate his movements in space and his qualities as a researcher.'

Ryumin's wife later disclosed her reactions to learning of her husband's Soyuz 35 assignment: 'I was rather shocked by his decision . . . Frankly, I didn't want this particular flight. I was worried about his health. From my vantage point, he hadn't yet rested properly after the previous mission that lasted nearly half a year, nor was he really back to full strength. I also felt that Valery would simply not have the time to prepare for the flight properly, and to go through all the training exercises.

I tried to persuade him to turn the offer down. He told me that he was the best person to cope with this assignment, that he knew the station better than anybody else, that he felt fully prepared for the job, and that long expeditions are a lot more interesting than short ones because one could accomplish a great deal more. I realised it was impossible to make him change his mind.'

Ryumin and Leonid Popov docked their spacecraft with the orbiting Salyut 6 space-station and carried out a long programme of research and experiments as well as playing host to the visiting cosmonauts from Hungary, Vietnam and Cuba.

On Ryumin's birthday, the second in a row he had spent in space, the communications session with his family was held from the Ryumin's new apartment. It was the first time he had seen it. They were preparing to move in soon after he returned from the Soyuz 32 mission, but he was then assigned to Soyuz 35.

He and Popov returned to Earth on 11 October 1980, after a flight lasting 184 days 20 hours and 12 minutes in space – thus setting a new duration record. He re-adapted so quickly that on the second day after his return to Earth, he was able to play tennis. He was subsequently awarded the title of Hero of the Soviet Union for the second time, and also received his third Order of Lenin. In addition, he was awarded the title of Hero of the Republic of Cuba, with Gold Star Medal and the Order of Playa-Giron, together with Vietnam's highest honours.

He acted as deputy Flight-Director during the mission of Kovalyonok and Savinykh in 1981, and during the missions of the Berezovoi/Lebedev and Kizim/Solovyov crews onboard Salyut 7 in 1983 and 1984 respectively, he served as Flight-Director. He continued in this role in 1985.

He is married, and his wife Natalya is an engineer and a graduate of the computer department of the Moscow Forestry Institute, where she first met Ryumin when they studied in the same group. They married in his second year at the Institute, and they now have two children, the eldest is a daughter, Vika, and the youngest a son, Vadim, born in 1972 or 1973.

Asked about her husband's job, Natalya Ryumin said 'It's very difficult being the wife of a cosmonaut because the profession is so exacting and demanding. Not just the flight but each training session, of which there are a great number, is a physical and emotional strain on the cosmonaut. It's not easy to be constantly aware that you have to help him ease the tension and relax. When Valery told me that he was about to become a member of the cosmonaut team, I cried and tried to dissuade him. Luckily for me, my mother was still living at the time. She was a wise woman, and she told me that I had no right to interfere.

And I didn't. However, if you should ask me now, after Valery has spent a year in space on the past two flights and has become famous, whether I want him to continue flying, I would say No. I try to keep my opinion to myself though, because I can see how fascinated he is with his work and how much satisfaction it gives him. So I've learned to hope for only one thing – for him to carry out all his plans successfully.'

Ryumin is a passionate hockey fan, enjoys tennis, and is said to be serious and efficient by nature. At 6ft 4in (1.93m) he is one of the tallest men in the cosmonaut team.

Lyakhov has described him as 'a very quiet, reasonable and business like man . . . an excellent engineer. Many people who see Ryumin on television get the impression that he is a gloomy, reserved man of few words. But his is merely the reaction of a person who is not used to being in the limelight. While he may look gloomy, he has an excellent sense of humour. He is a very good, attentive and responsive friend.'

During the Soyuz 35 mission, it was said that 'he likes a good joke, but he can be dead serious when he makes his report. He insists that flight control specialists give really exhaustive answers to his questions and he can be quite persistent on that score.'

His wife has said 'Work is Valery's life . . . He is an engineer, who himself worked on developing space technology, so he is well aware of how much labour has to be exerted to make each flight possible. I know that he would take a failure in orbit as a personal drama.'

Valery Ryumin was the 41st Soviet cosmonaut, and he has spent a total of 361 days 21 hours and 33 minutes in space.

Colonel Gennady Vasilyevich Sarafanov

Gennady Sarafanov was born on 1 January 1942, in Sinenkiye, a small village near Saratov, on the Volga. After finishing school, he joined the Soviet Army in 1959, and after graduating from the Balashov Higher Air Force Flying School, he served in various air units. He joined the CPSU in 1963.

He was selected to join the cosmonaut team in October/November 1965, at the age of 23. *It is thought that he may have trained for Salyut 2, but when that failed, he was re-assigned to Salyut 3. (see section entitled 'Salyut 2/Cosmos 557 Crewing' for more details.)*

His first spaceflight came as commander of Soyuz 15, launched on 26 August 1974, and the flight was widely expected to involve a month-long stay onboard the Salyut 3 space-station. However, a failure in the docking and rendezvous equipment prevented the docking taking place, and Sarafanov and Lev Demin were forced to return to Earth, landing in darkness after a mission lasting only 2 days and 12 minutes.

Soviet sources later stated that the cosmonauts had been testing an automatic docking system related to the development of an unmanned Soyuz designed to resupply manned Salyut stations with consumables. (i.e. the Progress spacecraft, which made its first appearance on 20 January 1978 with the launch of Progress 1 to the Salyut 6 space-station.)

Upon his return, Sarafanov was awarded the title of Hero of the Soviet Union and also received the Order of Lenin. He is said to be a keen philatelist, and is married. His wife, Tamara Vasilyevna, works in Star Town, and they have two children, Alexei, and Katya.

Gennady Sarafanov was the 31st Soviet cosmonaut.

Viktor Petrovich Savinykh

Viktor Savinykh was born on 7 March 1940, in the village of Berezkiny in Orichevsk Rayon, Kirov Oblast. His parents both worked on a farm co-operative, and his father was a veteran of World War Two, having fought at the Leningrad front.

He finished secondary school and then went to the Perm Technical School of Railway Transport, graduating in 1960. He then went to work as a team leader on the Sverdlovsk Railway in the Urals for several years before entering service in the Soviet Army. He joined the CPSU in 1963.

Savinykh served in the Army as a topographer, and after his national service was over, he stuck to his new calling and joined the Optical Instruments Faculty of the Moscow Institute of Geodetic Engineering, Aerial Photography and Cartography. He graduated in 1969 as a specialist in optical geodetic instruments.

Following graduation, he went to work in the Korolyov design bureau, developing equipment for outer space. He took part in designing optical instruments for orbital stations and transport spacecraft, and some of the instruments he helped to develop were installed on the Salyut 6 space-station. He took part in the ground control operations for manned spaceflights, including in 1975, the flight of Klimuk and Sevastyanov onboard Soyuz 18/Salyut 4. Savinykh was on the staff of the programme planning group and was responsible for onboard documentation.

In December 1978, he was selected to join the cosmonaut team. 'Nikolai Rukavishnikov, with whom I worked once, played a great part in deciding me to become a cosmonaut' he later said.

It is believed that he may have served as the flight-engineer on the support crew for Soyuz 35, launched on 9 April 1980. (see section entitled 'Salyut 6 Crewing' for more details.)

His next assignment was as the back-up to Gennady Strekalov, the research-cosmonaut onboard Soyuz T-3, launched on 27 November 1980. His first spaceflight came as flight-engineer onboard Soyuz T-4, launched on 12 March 1981. The mission was not originally planned and was required to provide a new long-duration crew onboard Salyut 6 to enable the Mongolian and Romanian Intercosmos missions to take place, the schedule having been disrupted by the Soyuz 33 failure. Savinykh began joint training with his commander, Vladimir Kovalyonok, in November 1980, following the completion of his Soyuz T-3 back-up assignment.

The two men spent nearly 74 days onboard Salyut 6 carrying out a series of experiments and also acting as hosts to the Mongolian and Romanian cosmonauts onboard Soyuz 39 and 40. They then returned to Earth on 26 May after a flight lasting 74 days 18 hours and 38 minutes.

Upon his return, Savinykh was awarded the title of Hero of the Soviet Union and also received the Order of Lenin. In making the flight, Savinykh had the distinction of not only being the first cosmonaut-geodesist, but also the 100th man in space.

Following the Soyuz T-4 mission, he defended a thesis on the subject of Geodesy, aerial photography and cartography using material from the mission and from his earlier work at the design bureau.

Savinykh's next assignment was as back-up to Alexander Serebrov, the flight-engineer onboard Soyuz T-7, launched on 19 August 1982.

It is also believed that when Serebrov was assigned to Soyuz T-8, launched on 20 April 1983, Savinykh again served as his back-up. In addition, it is thought that he may have served as the flight-engineer on the support crew for Soyuz T-9. He should then have been assigned to the back-up crew for the Soyuz T-10 mission, but following the 're-cycling' of Titov and Strekalov, he is believed to have been given a second support crew assignment on that mission instead. (see section entitled 'Salyut 7 Crewing' for more details.) However, when the mission suffered a launch abort, he was given the back-up assignment for the replacement Soyuz T-10, launched on 8 February 1984.

His next assignment was as back-up to Igor Volk, the research-cosmonaut on Soyuz T-12, launched on 17 July 1984.

When the crewing tables are examined, it appears that Savinykh was in training for a long-duration mission, and was temporarily sidetracked by the need to fly the Soyuz T-12 mission. Upon the completion of that mission, the original schedule was resumed, and he should then have flown on Soyuz T-13 with Vladimir Vasyutin and Alexander Volkov. (see section entitled 'Salyut 7 Crewing' for more details.)

However, when the Salyut suffered an onboard failure which left it without power, a repair mission became necessary. As a manual docking was required, Vladimir Dzhanibekov was given the assignment, and Savinykh was paired with him at the end of March 1985, as extensive onboard repairs were anticipated, and an experienced flight-engineer was needed.

The two men were launched onboard Soyuz T-13 on 6 June 1985, and linked up with Salyut 7 using a new manual docking system in a bid to rescue the lifeless station. They found the interior to be dark, freezing cold, and encrusted in ice. Working out of the Soyuz, which provided their only source of warmth, the two men carried out a series of repairs to the Salyut's electrical system, eventually returning it to a fully operational state. Their repair work salvaged the station, and enabled the Soviets to resume the planned flights to Salyut 7.

On 2 August, the two men carried out an EVA lasting 5 hours, during which they tested an upgraded spacesuit and installed an additional solar panel. They also installed a Medusa experiment and a device to collect dust from Halley's Comet.

Once the Salyut was declared operational, a new crew was launched to the station onboard Soyuz T-14. The two crews worked together for a week, and then Dzhanibekov returned to Earth with the Soyuz T-14 flight-engineer, Georgi Grechko, in the first partial replacement of a crew. Savinykh remained onboard the Salyut with the new arrivals, Vasyutin and A.Volkov, thus restoring the original Soyuz T-13 crew.

The three men settled down to a long programme of experiments and observations, and this included working with the Cosmos 1686 space-station module.

From later Soviet reports, it is clear that Savinykh fully expected to still be in orbit on 1 January 1986, and it seems likely that he was intended to set a new duration record. However, in November, Vladimir Vasyutin became ill and the mission had to be terminated early. The three men returned to Earth, Savinykh having spent 168 days 03 hours and 51 minutes in space.

He was subsequently awarded the title of Hero of the Soviet Union and also received the Order of Lenin, both for the second time.

Following his launch onboard Soyuz T-13, a Soviet report stated that 'As a back-up for other flights, Savinykh underwent a full training course for work in orbit four times during the last four years.' This would exclude the previously announced Soyuz T-3 back-up assignment, and three of the other assignments were later confirmed as the Soyuz T-7, T-10 and T-12 missions. The other assignment is believed to have been the back-up duty on Soyuz T-8 referred to earlier, but this has not yet been confirmed.

Savinykh is married, and he and his wife Lilya, a senior lecturer at the Physical Training Department of the Moscow Forest Engineering Institute, have a daughter, Valentina, born in 1968 or 1969. Savinykh's hobbies include tennis and swimming, and he has written a book entitled 'The Earth is Hoping and Waiting' published in 1984. When asked to describe him, Kovalyonok said 'He's calm by nature. It's impossible to upset him, which is very important for work in space. He's calm and controlled and does not make mountains out of molehills. He may be more energetic than is necessary when making decisions, but here we must compromise.' Vladimir Shatalov has described him as 'modest . . . a man of few words.'

During the training for Soyuz T-13, his fellow cosmonauts gave him the nickname 'Green Computer', because of the green overalls he wore during training sessions when he operated the computer complex while the others worked in the mock-up.

Viktor Savinykh was the 50th Soviet cosmonaut, and he has spent a total of 242 days 22 hours and 29 minutes in space.

Svetlana Yevgenyevna Savitskaya

Svetlana Savitskaya was born on 8 August 1948, in Moscow. Her father, Yevgeny Savitsky, is the deputy Commander of Soviet Air Defences, and a former World War Two air-ace. He was made an Air Marshall in 1961, and is a twice Hero of the Soviet Union. On her parents insistence, Savitskaya took up music, English and swimming in addition to school classes. She was fond of reading, and also went in for figure-skating and then running. It is said that as a young girl, she played with planes but never with dolls, and her childhood dream was to fly – especially high speed planes.

She decided to become a pilot, and as a 16 year old schoolgirl, turned up one day at an amateur flying club without either the knowledge or assistance of her father. She was rejected by the club as too young, but they did let her take a course in parachute jumping.

It was quite some time before her father accidently saw a parachute knife in her schoolbag, and the family learnt of her secret.

In eighteen jumps, she apparently became a master of parachute jumping, and while still only 17, she made three jumps from the stratosphere, each establishing a world record. She made a record sky-dive, jumping at an altitude of 14252m and performing a free-fall for nearly 14km before opening her parachute at a height of only 500m.

In 1965, upon graduation from the parachute course, Savitskaya already had to her credit 450 parachute jumps. At the age of 18, she was finally admitted to the piloting section of the flying club. She then enrolled in the aircraft-construction department at the Moscow Aviation Institute, and in her second year there, she made her first solo flight at the controls of a Yak-18.

She was dubbed 'Miss Sensation' by the British press in 1970 when she was included in the Soviet National Aerobatics team, and flew at the world aerobatics competition at Hullavington. At the time, she was still a student at the Moscow Aviation Institute. By then, her accrued flying time was just 300 hours – a trifle compared to her rivals – but she still became overall World Champion at the event.

In 1972, she graduated from the Institute as a mechanical-engineer, and remained an active competitor in flying tournaments. In the same year, she also graduated from the

Central Technical Flying School of the USSR Voluntary Society for the Promotion of the Army, Air Force and Navy (DOSAAF). She then worked as a flight-instructor at the Central Aviation club for several years. However, she was not content with this job, as her dream was to become a test-pilot. She began pressing for a place in a test-pilots training school, and learned to fly jets.

Eventually, in 1974, she was enrolled in a test-pilots school, first as a probation student, and eight months later as a a fully-fledged cadet. In parallel with studies at the school, she continued setting records on turbo-prop and jet planes, and in 1974, set eight world records for the rate of climb flying a supersonic plane. In the summer of 1975, she registered 2683.638 km/hr in a small mass-produced E-133 aircraft on a 15-25 km route – a new world record for a woman pilot. She set 18 world aviation records, and eleven of these records were still standing in 1982. Her latest records were set in the Spring of 1981, when she had already been selected to join the cosmonaut team. Flying a Yak-40 passenger plane with her friend Galina Korchuganova, she twice successfully lifted the maximum load to an altitude of more than 2km.

During her test-pilot career, she successfully tested more than 20 types of sports, passenger, subsonic and supersonic planes. She holds the qualification of 'Test-Pilot, 2nd Class,' and is a Merited Master of Sports of the Soviet Union. She has now made more than 500 parachute jumps, and has an accrued flying time of more than 2000 hours, including 400 hours in jets.

She joined the CPSU in 1975, and is currently a member of the Central Committee of the Komsomol – the Young Communist League (YCL.)

Having completed a course of training at the test-pilots school, she took up a test-pilots job at Alexander Yakovlev's design bureau in 1976.

She was selected to join the Soviet cosmonaut team in 1980, and according to Georgi Beregovoi, no concessions were made because she was a woman, and she underwent the same rigouress training as her male colleagues. 'It is no secret,' he said, 'that women are more accurate and pain-staking, especially in experiments that require fine and punctual work. In this situation, a woman cosmonaut may prove her worth to advantage.'

In a pre-launch interview, she said 'I have long been dreaming about space . . . I thought of it like any pilot who is striving to master new equipment, to achieve new records and higher speeds. I am a pilot and I regard a spaceship and a station as aircraft without wings. I am very eager to feel the fabulous speed of 28000 km/hr. A non-stop flight lasting a week is almost fantastic. Such things are not encountered in aviation. What is new for me is the experimental and research work. In a word, my profession as a pilot is being enriched by a new profession, that of cosmonaut-researcher.'

Her first spaceflight came as research-cosmonaut onboard Soyuz T-7, launched on 19 August 1982. The spacecraft docked with the Salyut 7 space-station, and Savitskaya, Leonid Popov and Alexander Serebrov joined the crew of Berezovoi and Lebedev already onboard. In making the flight, Savitskaya became the second woman in space, some 19 years after the first, Valentina Tereshkova, was launched onboard Vostok 6 on 16 June 1963. Observers pointed out that Savitskaya's flight came only 7 months before America was due to launch its first woman astronaut, Sally Ride, and it was therefore interpreted by some as a propaganda stunt.

Savitskaya has firm views about being a female cosmonaut. 'It's the job that counts,' she said, 'Whether it's done by a man or woman is irrelevent.' Anatoly Berezovoi apparently welcomed her onboard the Salyut with flowers, and then unwisely suggested that she cook a meal for the five cosmonauts. She quickly put him in his place by explaining that as host, that was his job.

The visiting crew spent nearly seven days carrying out a programme of experiments onboard the Salyut. Savitskaya's main task was to carry out extensive medical research to examine in greater detail the influence which weightlessness has on the female organism. The crew returned to Earth after a flight lasting 7 days 21 hours and 52 minutes.

Upon her return, Savitskaya was awarded the title of Hero of the Soviet Union and also received the Order of Lenin.

One year after her spaceflight, she said 'My dream is to fly into space again. I deal with technical training and have devoted a lot of time to the results of the Tavria biotechnological experiment, which I staged during my flight.

Unfortunately, I do not pilot planes as much as I would like any more, but my job has of course changed. I am a cosmonaut, not a pilot.'

Her second spaceflight came as flight-engineer onboard Soyuz T-12, launched on 17 July 1984, and she became the first woman to make a second spaceflight. Savitskaya, Vladimir Dzhanibekov and Igor Volk linked up with Salyut 7, and joined the crew of Kizim, Solovyov and Atkov already onboard.

During the mission, Dzhanibekov and Savitskaya carried out an EVA lasting 3 hours and 35 minutes. In doing so, Savitskaya became the first woman to perform an EVA, and although the spacesuit was specially modified, it was still not pliable enough to make movement easy, and it was therefore expected that she would lose a good deal of weight. In the course of the EVA, the two tested a new electron beam device, and Savitskaya used it to carry out a series of operations involving the cutting, welding and soldering of metal plates and the spraying on of a coating. During the experiments, Dzhanibekov carried out cine-photography and TV reporting. The two cosmonauts then changed places and the cycle of technological operations on other samples was repeated by Dzhanibekov. Following the successful EVA, the Soyuz T-12 crew conducted a series of experiments onboard the Salyut before returning to Earth after a flight lasting 11 days 19 hours and 14 minutes.

Savitskaya was subsequently awarded the title of Hero of the Soviet Union and also received the Order of Lenin, both for the second time.

She had first become acquainted with Igor Volk some years earlier, when they worked together as test-pilots. 'I met him earlier than anyone else in the cosmonauts detachment,' she said. 'When I was training at the Test Pilots School, he gave the graduation examination to our course.'

Following her second spaceflight, it was reported that she now works at a design bureau and participates in tests of space vehicles.

Savitskaya, known to her friends as Sveta, is married, and her husband, Viktor Khatkovsky, is an engineer-designer and pilot at the Ilyushin Aircraft design bureau. She is apparently the sister of a pilot, wife of a pilot, and daughter of a pilot. Her hobbies include sewing, knitting, and cookery, and she likes movies, drama, classical music, both Soviet and foreign – especially Tchaikovsky and Beethoven – and her favourite poet is Mikhail Lermentov, a 19th century Soviet poet.

She has been described as a slim and petite woman, of medium height, with dark chestnut hair, brown/hazel eyes, and is said to be fairly reserved in character.

When Leonid Popov was asked what qualities he appreciated most in Savitskaya, he said 'Her excellent knowledge of space technology – and her equally great courage.' Vladimir Dzhanibekov praised her high professional skill and persistence. Working outside the spacecraft, he said, demanded not only high skill but also physical strength, and she had shown 'stamina and strength comparable with those of a man.' 'She has a very complex and powerful character,' he continued, 'and is very purposeful and persistent in achieving an objective.'

Viktor Blagov, Deputy flight control chief, said 'Svetlana is a model of serious approach to studies and training, and she has unwittingly made the whole crew take matters seriously. Her flight commander and flight-engineer do not simply look after her, but obey her.'

According to several Soviet reports, Savitskaya is a difficult person to interview. She is very serious, and her expressions are 'outright and strong. She doesn't leave any doubt that she means what she says.' One journalist said 'It isn't all that easy to talk with her. She prefers accurate, business-like questions. Her answers are brief and clear. If she thinks that the question is far from the subject, she can correct the journalist and do it rather sharply too.'

Another reported 'She recognizes only one subject in her conversation – work. She gets indignant when she hears questions and references to herself as a woman. In general it is a bit difficult to interview her.'

By contrast, other reports have stated that she is 'charming and soft, a hospitable hostess who likes to make patterns and sew her own clothes when she has time to spare.'

Her father has said of her 'People say hers is a male disposition. It is hard to say . . . True, I've never seen her cry. She is strong-willed with tough nerves . . . whatever the situation she has never cried. She coils up like from a blow but never cries.'

Svetlana Savitskaya was the 53rd Soviet cosmonaut, and she has spent a total of 19 days 17 hours and 06 minutes in space.

Alexander Alexandrovich Serebrov

Alexander Serebrov was born on 15 February 1944 in Moscow. He graduated from the Moscow Physical-Engineering Institute in 1967, and worked as a researcher at one of the departments of the Institute for nine years.

In 1976, he joined the CPSU, and began work in the Petrov design bureau. He soon showed himself to be a highly-qualified and resourceful worker capable of solving complicated engineering and scientific tasks independently. At the design bureau, he took part in the design and testing of spacecraft, including the development of the Soyuz T, and whilst there, met Gennady Strekalov, another future cosmonaut.

He was selected to join the cosmonaut team in 1978, and it is believed that he continued to be involved in Soyuz T development work. His first spaceflight came as flight-engineer onboard Soyuz T-7, launched on 19 August 1982. The spacecraft docked with the Salyut 7 space-station, and Serebrov, Leonid Popov and Svetlana Savitskaya joined the crew of Berezovoi and Lebedev already onboard. The visiting crew spent nearly seven days carrying out a programme of experiments before returning to Earth after a flight lasting 7 days 21 hours and 52 minutes.

Upon his return, Serebrov was awarded the title of Hero of the Soviet Union and also received the Order of Lenin.

A problem arose with Salyut 7's solar panels, resulting in a reduced flow of electrical power. To overcome this serious problem, a repair crew of Vladimir Titov and Gennady Strekalov were assigned to train for a mission launched specifically to repair the faulty Salyut 7 solar array. Serebrov was also assigned as research-cosmonaut for the mission, Soyuz T-8, launched on 20 April 1983.

The three men were supposed to dock with Salyut 7, but once in orbit, the Soyuz rendezvous radar antenna failed to deploy properly. Several attitude control manoeuvres at high rates were made but failed to swing the boom out. (The post-flight inquiry later discovered that the whole of the antenna had been torn off when the Soyuz payload shroud separated.) The crew received permission from FCC to attempt a rendezvous using only an optical sight and ground radar inputs for guidance. However, the final approach had to be made in darkness, and Titov believed that the closing speed was too great. He therefore attempted a braking manoeuvre, but felt that the two spacecraft were still closing too fast. He therefore aborted the rendezvous to avoid a crash, and no further attempts were made before the three men returned to Earth after a flight lasting 2 days and 18 minutes.

Upon his return, Serebrov did not receive any of the usual Soviet awards, indicating official dissatisfaction with the failure of the mission.

In September 1985, he acted as a commentator on Soviet TV during the Soyuz T-14 docking, and it is thought that he may have been a member of the back-up crew. If this is correct, he was probably also the back-up for the original Soyuz T-13 mission, before it was cancelled.

He is a physicist and designer, and is said to be a specialist on mass exchange and 'everything connected with heat transfer, from the heating of the station by the Sun to technological ovens.' He holds a degree as a Candidate of Technical Sciences.

Serebrov – a tall and athletically-built man – is known to his friends as Sasha, and his interests include fast cars. He is married, and he and his wife Ekaterina have two children, including a son, Kirill, born in 1969 or 1970.

Alexander Serebrov was the 52nd Soviet cosmonaut, and he has spent a total of 9 days 22 hours and 10 minutes in space.

Vitaly Ivanovich Sevastyanov

Vitaly Sevastyanov was born on 8 July 1935, in Krasnouralsk in the Sverdlovsk Region, the son of a taxi-driver. He went to the Ostrovsky school in Sochi, a seaside resort on the Black Sea coast, where he spent much of his childhood. During his student holidays, he worked as a deckhand on small pleasure boats cruising between Sochi and Khosta. In 1959, he graduated from the Moscow Aviation Institute, and began work in a design office. While working there, he took a post-graduate course at the Aviation Institute, and won a Candidate of Technical Sciences degree in 1965. He joined the CPSU in 1963.

He first became involved with spaceflight during his post-graduate studies, when he began giving lectures on rocketry and celestial mechanics to the cosmonaut classes, and he was in fact one of the first teachers at the cosmonaut training centre. He was involved in the Vostok 1 mission, and later recalled 'I had the good luck of listening to the broadcast of the preparations for the launch and the blast-off itself while at the flight control centre in Moscow as a representative of the design office.'

He was selected to join the cosmonaut team in 1966, and arrived in Star Town in November of that year, although he did not start training until 8 January 1967. In mid-1967, he was assigned to the Zond training group as training manager, but was replaced by Pavel Belyayev in late 1967 when he was re-assigned to the Soyuz programme. *According to one report, he was assigned as flight-engineer for the first test-flight of the Soviet Lunar Module in lunar orbit (see section entitled 'The Zond programme' for more details.) However, this report appears to contradict Sevastyanov's own version of events i.e. the switch from the lunar programme to the Soyuz programme.*

His first assignment was as back-up flight-engineer for the Soyuz 6, 7 and 8 flights, all launched in October 1969. His first spaceflight came as flight-engineer aboard Soyuz 9, launched on 1 June 1970. The flight was a long-duration one to test the effects of prolonged weightlessness on Man. After making a successful re-entry and touchdown, Sevastyanov and his commander, Andrian Nikolayev, had to be carried from the descent module. Their record flight of 17 days 16 hours and 59 minutes onboard the cramped Soyuz spacecraft had left them weak, and they were unable to walk or sleep properly for the next four or five days. Upon his return, Sevastyanov was awarded the title of Hero of the Soviet Union, and also received the Order of Lenin.

In March 1972, a Western news agency reported that Sevastyanov was going to make a flight onboard a new Salyut in May of that year. The report is a strange one, particularly as it indicated that he was intended to fly with one of two rookie cosmonauts – Vinogradov or Voronov. (see section entitled 'Salyut 2/Cosmos 557 Crewing' for more details.)

Sevastyanov was teamed with Pyotr Klimuk in early 1972, and later flew a long-duration mission with him in 1975. It is puzzling therefore that Klimuk flew with Valentin Lebedev on Soyuz 13, launched on 18 December 1973, particularly as Lebedev had only been a cosmonaut for a year, and had not performed any back-up duties. In addition, he is known to have begun his training with Vladimir Kovalyonok. It is therefore possible to speculate that Sevastyanov was originally assigned to fly with Klimuk and that he was replaced by his back-up – Lebedev – at the last-minute when he became disqualified for some reason. It is interesting to note that the Soyuz 13 back-up crew has never been identified.

It is believed that Sevastyanov's next assignment was as the back-up to Georgi Grechko, the flight-engineer onboard Soyuz 17, launched on 10 January 1975. (see section entitled 'Salyut 4 Crewing' for more details.)

He then served as the back-up to Oleg Makarov, the flight-engineer onboard Soyuz 18, launched on 5 April 1975. Following a booster malfunction during launch, the mission was aborted, and was subsequently designated the 'April 5 anomaly.' Sevastyanov then flew on the replacement mission, termed Soyuz 18, launched on 24 May 1975.

The crew of Sevastyanov and Pyotr Klimuk linked up with the Salyut 4 space-station and spent two months onboard, carrying out many experiments and Earth observations. They then left the station and returned to Earth after a flight lasting 62 days 23 hours and 20 minutes.

Upon his return, Sevastyanov was awarded the title of Hero of the Soviet Union, and also received the Order of Lenin, both for the second time.

In September 1977, he attended the 28th Congress of the International Astronautical Federation, held in Prague, and took part in the commission on applied aspects of space research.

In the period 1977-1979, he is known to have been involved in the ground control of the Soyuz 26, 28, 30 and 32 missions.

Sevastyanov has become a leading expert on the Soyuz programme, and is also the author of more than twenty scientific papers. He devotes much of his time to public affairs, and is Chairman of the Soviet Chess Federation and Vice-President of the Soviet-Finland Friendship Society. He is a member of the Union of Journalists of the Soviet Union, and is a commentator on the Soviet Central Television series 'Man, the Earth and the Universe' – a popular science programme. By making the Soyuz 18 flight, he became the first journalist in space. He has been variously described as lively, likeable and cheerful, quick and emotional, 'always smiling, ready at any time to return a joke with a joke.'

He is married, and his wife Alevtina works for the Novosti Press Agency, having graduated from the philology faculty of Leningrad State University. They have a daughter Natasha. Sevastyanov's parents, Ivan and Tatyana Sevastyanov are both pensioners.

Vitaly Sevastyanov was the 22nd Soviet cosmonaut, and he has spent a total of 80 days 16 hours and 19 minutes in space.

Major Rakesh Sharma

Rakesh Sharma was born on 13 January 1949, in Patiala, India. He graduated from the Nizam College in Hyderabad, and in 1966, enrolled as an Air Force cadet at India's National Defence Academy, at Khadakvasla. Upon graduation, he was commissioned into the Indian Air Force. He served in a number of different operational squadrons, including a MiG squadron in the Bangladesh war in 1971, before taking a test-pilots course, after which he tested a number of different planes.

In June 1979, the Soviet Union offered India the chance to fly a cosmonaut onboard a JointSoviet-Indian spaceflight, along the same lines as the agreement to fly non-Soviet cosmonauts from member countries of the Intercosmos programme. After further discussions, the offer was accepted by Indian Prime Minister Gandhi in March 1981, and India began a cosmonaut selection programme. The two finalists were Sharma and Ravish Malhotra and they were sent to begin their training at the Yuri Gagarin Cosmonaut Training Centre, arriving on 19 September 1982. At the time of his selection, Sharma had 1600 flying hours to his credit.

In June 1983 the two men were taken to the Black Sea port of Fedosia to practice splashdown procedures. During these exercises, photographs were released showing Malhotra training with Yuri Malyshev and Nikolai Rukavishnikov, and Sharma with Anatoli Berezovoi and Georgi Grechko. It was assumed that these were the crews which would make the actual flight to Salyut 7, with the Malhotra crew as the prime choice, as Malyshev had originally been scheduled to command the joint Soviet-French mission before he was stood down.

In early July the two cosmonauts arrived back in India for a months holiday, and upon their return to Star Town in August, were paired with their respective commanders and flight-engineers.

Finally, on 6 October 1983, the crews for the joint flight were officially named. The prime crew consisted of Malyshev, Rukavishnikov (subsequently replaced by Gennady Strekalov) and Sharma; the reserve crew was Berezovoi, Grechko and Malhotra. Therefore, for some reason Malhotra had been replaced as prime by Sharma. According to the Indian press it was because Sharma had a better command of the Russian language. He also speaks Punjabi, Hindi and Telagu, and during the onboard press conference, he demonstrated his linguistic abilities by answering an Indian reporter's question in fluent English.

He made his spaceflight as research-cosmonaut onboard Soyuz T-11, launched on 3 April 1984. Sharma, Malyshev and Strekalov docked with the orbiting Salyut 7 space-station and joined the crew of Kizim, Solovyov and Atkov already onboard. The three men carried out a programme of joint Soviet-Indian experiments and observations, including a unique experiment involving a series of Yoga exercises to see whether they could help to reduce sickness by improving blood circulation and body balance.

The three men returned to Earth after a mission lasting 7 days 21 hours and 41 minutes.

Following the successful completion of the mission, Sharma was awarded the title of Hero of the Soviet Union and also received the Order of Lenin, together with the Indian Ashoka Chakra Order. In addition, a street in the village of Znemenovka, in the Ukraine was named Rakesh Sharma Street in his honour.

After the flight, he said 'What are my plans? I will go back to my previous job as a test-pilot. My knowledge and experience of the conquest of space may also come in handy.' Grechko once described Sharma as 'capable, modest and good-natured.' His hobbies include skiing, which he learned during his two winters at Star Town.

Sharma is married, and he and his wife, Madhu, a design artist, have a son, Kapil, born in 1974 or 1975. Their daughter, Mansi, died in 1983, aged 6, while the whole family were living in Star Town. Sharma's father, Devandranath Sharma, and his mother, Tripta Sharma, live in Hyderabad, where Sharma spent his childhood.

Rakesh Sharma was the 11th Intercosmos cosmonaut.

Lieutenant-General Vladimir Alexandrovich Shatalov

Vladimir Shatalov was born on 8 December 1927, in Petropavlovsk, a city in Northern Kazakhstan. His father was an engine driver on the Soviet railway who was awarded the title of Hero of Socialist Labour for his selfless work. Shatalov spent his childhood and early school years in Leningrad, and then attended the special Air Force school in Lipetsk upon the recommendation of model airplane experts, who had been impressed with the models he had constructed. He then went to the Kachinsk Air Force Pilots College, graduating in 1949 with honours, and became a flight-instructor. In 1953, the same year in which he joined the CPSU, he went to the Red Banner (now Gagarin) Air Force Academy, and graduated with distinction in 1956.

He served in the Soviet Air Force as a deputy squadron-leader, then as first deputy regimental commander, and eventually as an Air Force inspector. He was selected to join the cosmonaut team in 1963, and reported to Star Town on 11 January.

It is said that he soon became known on the cosmonaut team as an excellent engineer, and a cool-headed, tireless investigator. His first assignment was as commander of the support crew for Voskhod 1, launched on 12 October 1964.

It has been revealed that a *third* mission in the Voskhod series was planned for the early summer of 1966, involving a two-man team of cosmonauts in a long-duration mission. A training group was formed, probably around April 1965, but the mission was cancelled in early 1966.

The four cosmonauts since named as being part of this group were Shatalov, Beregovoi, Volynov and Shonin. It is thought that the prime crew would have been Volynov and Shonin, with Beregovoi and Shatalov as the back-ups. Following cancellation of the remaining Voskhod missions, the four men were re-assigned to the new Soyuz programme.

Shatalov's next assignment was as back-up to Georgi Beregovoi, the commander of Soyuz 3, launched on 26 October 1968, and his first spaceflight came as pilot of Soyuz 4, launched on 14 January 1969. After one day in orbit, he was joined by Soyuz 5, and Shatalov manually docked with the other spacecraft in the world's first docking of two manned spacecraft. Yevgeny Khrunov and Alexei Yeliseyev then carried out an EVA to transfer from Soyuz 5 to Soyuz 4, and returned to Earth with Shatalov, the Soyuz 4 descent module touching down after a flight lasting 2 days 23 hours and 21 minutes.

Upon his return, Shatalov was awarded the title of Hero of the Soviet Union and also received the Order of Lenin. His next spaceflight came as commander of Soyuz 8, launched on 13 October 1969. Shatalov and Alexei Yeliseyev carried out rendezvous manoeuvres and other experiments in a joint flight with the Soyuz 6 and 7 spacecraft. During the joint operations, Shatalov was in overall command of all three spacecraft. Soyuz 8 returned to Earth after a flight lasting 4 days 22 hours and 51 minutes.

Shatalov was subsequently awarded the title of Hero of the Soviet Union and also received the Order of Lenin, both for the second time.

In a group photograph of the Soyuz 6, 7 and 8 prime and back-up crews published in Georgi Shonin's book, 'The Very First Ones', an unknown man bearing a strong resemblance to the 1960 cosmonaut known as Dmitri is shown standing next to Yeliseyev. Although Shatalov also appears in the photograph, he is sitting at the front with Beregovoi, Kamanin and two Air Force Generals, whereas the prime and back-up cosmonauts are all standing in the back row.

From the photograph, it can be deduced that Dmitri was the original commander of Soyuz 8, but was replaced by Shatalov at a relatively late stage when he became disqualified on medical grounds. Dmitri left the cosmonaut team shortly afterwards.

If this speculation is correct, it would mean that Shatalov was never intended to fly on Soyuz 8. It would also help to explain why he flew three missions almost in succession i.e. Soyuz 4, 8 and 10.

His third spaceflight came as commander of Soyuz 10, launched on 23 April 1971. *It is possible that Shatalov was originally intended to fly on the next mission, Soyuz 11, but the crew was switched due to the illness of the original Soyuz 10 crew. If this is correct, Shatalov may have then served as the back-up commander for Soyuz 11. (See section entitled 'Salyut 1 Crewing' for more details.)*

The Soyuz 10 spacecraft docked with the Salyut 1 space-station – the first time such an operation had taken place – but undocked after only 5 hours and 30 minutes, without the crew entering the station. The Soviets have stated that it was never their intention for the station to be manned during this initial step, but in reality it would seem likely that there was actually some problem with the transfer hatch, preventing the crew from entering the Salyut. Shatalov, Nikolai Rukavishnikov and Alexei Yeliseyev returned to Earth after a flight lasting 1 day 23 hours and 46 minutes. Shatalov was subsequently awarded his third Order of Lenin, and was promoted to Major-General.

In September/October 1971, Shatalov was appointed Director of Cosmonaut Training at the Yuri Gagarin Cosmonaut Training Centre, the position he currently holds, taking over from General Kamanin. He is President of the Soviet – Cuban Friendship Society, and is a Deputy to the Supreme Soviet, having been a delegate to the 24th Congress. In addition to the awards and honours already listed, he has also been awarded the IAF Vladimir Komarov diploma, and the Yuri Gagarin Gold Medal. In 1972, he was awarded a Candidate of Technical Sciences degree. His favourite hobby is underwater exploration, and he is the author of a book 'Trudnye Dorogi Kosmosa' (Hard Roads Into Space.)

Following the completion of the joint Soviet-Vietnamese Soyuz 37 mission, launched on 23 July 1980, Shatalov was awarded the title of Hero of Labour of the Socialist Republic of Vietnam with the Order of Ho Chi Minh, 1st Class, for his part in the training of the

Vietnamese cosmonauts. In late 1982, he was appointed Chief of Space Navigation, which was seen as a response to the formation of the US Space Command at Peterson AFB., California. In a book published in 1984, it was claimed that Shatalov was one of the top people in GRU (Soviet Military Intelligence) and is their chief contact at the Yuri Gagarin Cosmonaut Training Centre.

Shatalov is said to be an extremely well-read man, and when putting forward an opinion, keeps his comments brief and to the point. He is married, and his wife Muza Andreevna is a scientist, specializing in agriculture. They met when she was taking her post-graduate course at the Novocherkassy Agriculture Institute where she graduated with distinction as a Candidate of Science (Agriculture). After graduating, she was sent to work at a Leningrad research institute, and after a time she asked to be transferred to a plant breeding station in the Crimea. Shatalov was to say later ' Of the 17 years of our married life, it's only some 5 years since we've been living under the same roof. We're as good as newlyweds ! '. They have two children, Elena, born in 1958, and Igor – a graduate of the Moscow Aviation Institute – born in 1952. Igor has not become a pilot however, but works in Leningrad.

Vladimir Shatalov was the 13th Soviet cosmonaut, and he has spent a total of 9 days 21 hours and 58 minutes in space.

Major-General Georgi Stepanovich Shonin

Georgi Shonin was born on 3 August 1935, in Rovenki, in the Lugansk Region of the Ukraine, and his childhood years were spent in Balta, not far from Odessa. His father was killed in World War 2, and following secondary school, Shonin went to the Odessa Special Aviation School. Upon graduation, he went to the Naval Air Force College, and then served as a pilot in the Baltic and Northern fleets. He joined the CPSU in 1957.

He was selected to join the cosmonaut team in 1960, and his first assignment came when he and Grigori both served as support cosmonauts for Vostok 3, launched on 11 August 1962. *Shonin was also involved in some way with the flight of Voskhod 2, launched on 18 March 1965, probably as a member of the support crew.*

It has been revealed that a *third* mission in the Voskhod series was planned for the early summer of 1966, involving a two-man team of cosmonauts in a long-duration mission. A training group was formed, probably around April 1965, but the mission was cancelled in early 1966.

The four cosmonauts since named as being part of this group were Shonin, Beregovoi, Volynov and Shatalov. It is thought that the prime crew would have been Volynov and Shonin, with Beregovoi and Shatalov as the back-ups. Following cancellation of the remaining Voskhod missions, the four men were re-assigned to the new Soyuz programme.

As part of his training, Shonin attended the Zhukovsky Military Air Academy and graduated in 1968. His next assignment was as the back-up to Boris Volynov, the commander of Soyuz 5, launched on 15 January 1969, and his first spaceflight came as the commander of Soyuz 6, launched on 11 October 1969. Shonin and Valeri Kubasov carried out joint manoeuvering experiments with the Soyuz 7 and 8 spacecraft, and in addition, Kubasov carried out the world's first space-welding and metals smelting experiments. The two men returned to Earth after a flight lasting 4 days 22 hours and 43 minutes.

Upon his return, Shonin was awarded the title of Hero of the Soviet Union and also received the Order of Lenin.

According to one Western report, a heavy Zond-Proton countdown was conducted on 3 and 4 July 1969, and the cosmonaut crew present at Baikonur was thought to consist of Filipchenko, Kubasov and Shonin. Their launch was cancelled when the G-vehicle failed shortly after lift-off. (see section entitled 'The Zond programme' for more details.)

In July 1975, Shonin was the Chief Soviet CapCom assigned to the ASTP mission, and in 1976, he was promoted to Major-General. He is the author of a book, 'Samiye Pyerviye' ('The Very First Ones'), published in 1977, and in 1978, he was awarded a Candidate of Technical Sciences degree. In 1980, he was described as Head of the Training department at the Yuri Gagarin Cosmonaut Training Centre, a position which he still holds. In 1984, he revealed that he was involved in development work on the Soviet shuttle.

He is married, and he and his wife Lydia have two children, Andrei and Nina. Shonin, who has been described as 'dark-eyed, reserved and tactful', is said to be an admirer of the works of Antoine de Saint-Exupéry, the French pilot and author.

Georgi Shonin was the 17th Soviet cosmonaut.

Vladimir Alexeievich Solovyov

Vladimir Solovyov was born on 11 November 1946, in Moscow. His father was an aviation research engineer and his mother a higher mathematics teacher. Upon finishing school, he enrolled at the E.N. Bauman Moscow Higher Technical School, and after graduating in 1970, he started work in Sergei Korolyov's design bureau.

He joined the CPSU in 1977, and in the same year, applied to join the cosmonaut team. He was accepted in 1978.

His first assignment came in October 1981, when he was named as back-up to Alexander Ivanchenkov, the flight-engineer for the joint Soviet-French mission, Soyuz T-6, launched on 24 June 1982. For his part in the preparations for the mission, he was awarded the French Order of the Legion of Honour.

In 1982, Solovyov joined the special Salyut training group preparing for long-duration missions.

It is thought that his next assignment was as the flight-engineer on the support crew for Soyuz T-8, launched on 20 April 1983. It is believed that he and Leonid Kizim were then given the back-up assignment for Soyuz T-9, launched on 27 June 1983. During Lyakhov and Alexandrov's 2 EVA's from Salyut 7 in November 1983, Kizim and Solovyov conducted similar activities in the water tank at the Yuri Gagarin Cosmonaut Training Centre.

The two men should have been the prime crew on the next mission, but because of the 're-cycling' of Titov and Strekalov, they again served as the back-up crew. (See section entitled 'Salyut 7 Crewing' for more details.)

Solovyov's next assignment therefore came as back-up to Gennady Strekalov, the flight-engineer onboard Soyuz T-10, launched on 27 September 1983. The mission was aborted following a launch pad fire, and Solovyov and Kizim were re-assigned as prime crew for the replacement flight. The aborted mission was never given a designation, and Solovyov, Kizim and Oleg Atkov were launched onboard the new Soyuz T-10 on 8 February 1984.

The three men docked their spacecraft with the orbiting Salyut 7 space-station and carried out a long programme of research and experiments as well as playing host to the joint Soviet-Indian spaceflight Soyuz T-11, and the visiting Soyuz T-12.

A major task carried out by Solovyov and Kizim was a series of six EVA's during which they completed extensive and complicated repairs to the Salyut.

The first four EVA's were concerned with the fuel system. The two men opened the side of Salyut 7 and replaced an oxidizer line which had leaked, damaging the Salyut propulsion system. They also installed a new but separate parallel propellant line in the area where the failure occurred.

The four EVA's took place in the period 23 April – 4 May, and lasted 4 hours and 15 minutes; 4 hours and 56 minutes; 2 hours and 45 minutes; and 2 hours and 45 minutes respectively.

A fifth EVA lasting 3 hours and 5 minutes was conducted on 18 May, during which two extra solar panels were installed on one of Salyut 7's three main arrays.

One more EVA was scheduled to complete the repairs to the fuel system. Vladimir Dzhanibekov had undergone special training in the water tank at the Yuri Gagarin Cosmonaut Training Centre and had practised with the tools and procedures necessary to compress and seal the ruptured fuel line. It was originally intended that he would carry out the repair himself during the Soyuz T-12 visit. However, Kizim and Solovyov, having carried out the four preparatory EVA's, asked to complete the job themselves. Therefore, Dzhanibekov delivered a section of pipeline and a video film of his water tank repair procedure, and showed how the repairs should be carried out.

Solovyov and Kizim conducted the final EVA lasting 5 hours on 8 August, during which they compressed and sealed of the fuel line with the aid of a special tool. They had spent a total of 22 hours and 46 minutes conducting the six EVA's.

The three men resumed their normal programme of research for a further two months, before returning to Earth after a flight lasting 236 days 22 hours and 50 minutes – thus setting a new duration record.

Upon his return, Solovyov was awarded the title of Hero of the Soviet Union and also received the Order of Lenin. For his part in the joint Soviet-Indian mission, he was also awarded the Indian Order of Kirti Chakra.

In 1985, he served as deputy flight leader for the Soyuz T-13 mission.

Solovyov, known to his friends as Volodya, is married, and he and his wife Elena Ioureva, a mechanical engineer, have a son, Sergei, born in 1970, and a daughter, Maria, born in 1981. His hobbies include travel, hiking and sailing.

He has worked as a shift flight director, and is reported to be a specialist on power plants and on refueling the Salyut OMS (Orbital Manoeuvring System.)

Vladimir Solovyov was the 56th Soviet cosmonaut.

Gennady Mikhailovich Strekalov

Gennady Strekalov was born on 28 October 1940, in the town of Mytishchi, near Moscow. His father was killed in action in 1945 during the liberation of Poland. After finishing school, Strekalov worked in a factory for two years, and as an apprentice coppersmith, he was present for the fabrication of the Sputnik 1 spacecraft. He then entered the E.N. Bauman Moscow Higher Technical School, and upon graduation in 1965, he went to work in Sergei Korolyov's design bureau. He earned the reputation of being an innovative and competent engineer, with vast knowledge, and plenty of initiative. He proved to be a good designer, having two inventions patented, and he worked on the design and testing of new space vehicles, including the Soyuz spacecraft. He joined the CPSU in 1972.

He was selected to join the cosmonaut team in 1973, and his first assignment was as back-up to Vladimir Aksyonov, the flight-engineer onboard Soyuz 22, launched on 15 September 1976.

He began training as a flight-engineer for a Salyut mission in October 1978, and his next assignment was as the back-up flight-engineer for Soyuz 35, launched on 9 April 1980. *It is thought that he may also have served on the support crew for Soyuz 32 – see section entitled 'Salyut 6 Crewing' for more details.)*

His first spaceflight came as research-cosmonaut onboard Soyuz T-3, launched on 27 November 1980. He replaced Konstantin Feoktistov, who, it appears, had to be dropped from the mission at a very late stage, presumably for health reasons. The flight was the first test flight of the three man version of the Soyuz T spacecraft. The role Strekalov played – that of research-cosmonaut – was the new position afforded by the third seat. The spacecraft docked with the Salyut 6 space-station, and Strekalov, Leonid Kizim and Oleg Makarov spent 10 days carrying out repairs to the station before returning to Earth after a mission lasting 12 days 19 hours and 8 minutes.

Upon his return, Strekalov was awarded the title of Hero of the Soviet Union and also received the Order of Lenin. His next assignments were as back-up to Viktor Savinykh, the flight-engineer onboard Soyuz T-4, launched on 12 March 1981, and then as back-up to Valentin Lebedev, the flight-engineer onboard Soyuz T-5, launched on 13 May 1982.

A problem arose with Salyut 7's solar panels, resulting in a reduced flow of electrical power. This seriously affected the environmental control system, resulting in cold, damp and uncomfortable conditions which threatened to make the station uninhabitable, and also posed a threat to the continued function of the station's electronics and other systems.

To overcome this serious problem, a repair crew of Strekalov and Vladimir Titov were assigned to train for a mission launched specifically to repair the faulty Salyut 7 solar array. Strekalov therefore made his second spaceflight as flight-engineer onboard Soyuz T-8, launched on 20 April 1983. Strekalov, Titov, and Alexander Serebrov were supposed to dock with Salyut 7, but once in orbit, the Soyuz rendezvous radar antenna failed to deploy properly. Several attitude control manoeuvres at high rates were made but failed to swing the boom out. (The post-flight inquiry later discovered that the whole of the antenna had been torn off when the Soyuz payload shroud separated.) The crew received permission from FCC to attempt a rendezvous using only an optical sight and ground radar inputs for guidance. However, the final approach had to be made in darkness, and Titov believed that the closing speed was too great. He therefore attempted a braking manoeuvre, but felt that the two spacecraft were still closing too fast. He therefore aborted the rendezvous to avoid a crash, and no further attempts were made before the three men returned to Earth after a flight lasting 2 days and 18 minutes.

Upon his return, Strekalov did not receive any of the usual Soviet awards, indicating official dissatisfaction with the failure of the mission. He and Titov were then apparently 'recycled' and flew on the next available mission in a second attempt to carry out the repairs that they had been trained for.

The two men were to have been launched onboard what should have been Soyuz T-10 on 27 September 1983. However, a fire started in the base of the launch vehicle when a valve in the propellant line failed to close at T-90 seconds. This malfunction ignited a large fire which engulfed the rocket, and the automatic abort sequence failed as the wires involved burnt through. The launch controllers were forced to manually abort the mission by sending radio commands from the launch blockhouse.

The Soyuz descent module was pulled clear by the launch escape system, and the crew landed safely some 2.5 miles (4 km) from the rocket, which apparently exploded seconds after the Soyuz separated. The two men were given a medical check-up, but had sustained no injuries. Had the mission succeeded, it appears that Titov and Strekalov would have become the new resident crew, thus achieving the first handover of a Salyut from one crew to another.

Strekalov's third spaceflight came as flight-engineer onboard Soyuz T-11, launched on 3 April 1984. He replaced Nikolai Rukavishnikov, who had to be dropped from the mission at a late stage, after being taken ill. Strekalov, Yuri Malyshev and Rakesh Sharma docked with the orbiting Salyut 7 space-station and joined the crew of Kizim, Solovyov and Atkov already onboard. The three men carried out a programme of joint Soviet-Indian experiments and observations, before returning to Earth after a flight lasting 7 days 21 hours and 41 minutes. Strekalov was subsequently awarded the title of Hero of the Soviet Union and also received the Order of Lenin, both for the second time. He also received the Order of Ashoka Chakra, India's supreme award.

Strekalov is, according to Soviet sources, one of the engineers responsible for the designing and ground testing of the Soyuz T spacecraft, and for the development of the Salyut stations. *In September 1985, he acted as a commentator on Soviet TV during the Soyuz T-14 docking, and it is thought that he may have been a member of the back-up crew. If this is correct, he was probably also the back-up for the original Soyuz T-13 mission, before it was cancelled.*

He is married, and he and his wife Lydia have two daughters – one born in 1975 or 1976, and the other – Tatyana – born in 1974 or 1975. Strekalov has been described as 'stern, reserved and slightly ironic about himself. Very polite and friendly. Other people feel at ease in his company.'

Gennady Strekalov was the 49th Soviet cosmonaut, and he has spent a total of 22 days 17 hours and 07 minutes in space (excluding the Soyuz launch abort.)

Colonel Arnaldo Tamayo-Méndez

Arnaldo Tamayo-Méndez was born on 29 January 1942 in the city of Guantanamo, Cuba. He was orphaned at the age of one, and was adopted by foster parents – Rafael Tamaya, a worker, and Esperanza Méndez, a housewife. He began work at the age of 13 as a shoe-shine boy and vegetable seller, and his first regularly paid job was as an apprentice carpenter. Following the Cuban Revolution, he joined the Association of Young Rebels, and then the Revolutionary Work Youth Brigades. This enabled him to continue his studying at a technical college.

In December 1960, he enrolled in a course for aviation technicians, and then decided to become a pilot. In April 1961, he left for an intensive one year course at an Air Force Academy in the Soviet Union. He returned to Cuba in May 1962, and was assigned to the Playa Giron Brigade of the Guard. He was later transferred to a pursuit group in Cuba's central region, where he remained until he was selected for cosmonaut training. In 1963, he was promoted to Lieutenant and placed in charge of an Air Unit's combat training, and in 1967 he was made squadron head. In 1967/68, he was sent to Vietnam to study the conflict with the U.S.

In 1968, Tamayo-Méndez was promoted to First Lieutenant, and made a military pilot 1st Class. A year later, he was admitted to the General Maximo Gomez Basic College of the Revolutionary Armed Forces where he took advanced Command and Staff Courses. He joined the Communist Party of Cuba in 1968. From 1971-75, he served as staff chief of a fighter brigade, while continuing to carry out his duties as a pilot and participating in scheduled flights. In 1972, he was promoted to Sub-Captain, then in 1973 to Sergeant-Major, and in 1975 to Major. Also in 1975, he was named as alternate chief (2nd in command) of a large air unit. In 1976, he was promoted to Lieutenant-Colonel. He has logged over 1400 hours in the air.

Following the agreement to fly non-Soviet cosmonauts from member countries of the Intercosmos programme onboard Soyuz and Salyut spacecraft, Cuba began a cosmonaut selection programme. A large number of Cuban Air Force pilots were chosen for the aptitude tests and medical examinations. Nine finalists were chosen, of which four were sent to the Soviet Union in January 1978. They spent the last ten days of January and all of February undergoing numerous tests before returning to Cuba.

The two finalists were Tamayo-Méndez and José Armando Lopez-Falcon, who learnt of their selection on 1 March 1978 and began their training at the Yuri Gagarin Cosmonaut Training Centre on 22 March 1978.

In September 1979, Tamayo-Méndez was teamed up with Yuri Romanenko in preparation for the joint flight. He eventually flew as research-cosmonaut onboard Soyuz 38 launched on 18 September 1980. The two men docked with the orbiting Salyut 6 space-station, and joined the crew of Popov and Ryumin already onboard. They then carried out a joint programme of Soviet-Cuban experiments and observations before returning to Earth after a mission lasting 7 days 20 hours and 43 minutes.

Upon his return, Tamayo-Méndez was awarded the title of Hero of the Soviet Union and also received the Order of Lenin. In addition, he was awarded the title of Hero of the Republic of Cuba, with Gold Star Medal and Order of Playa Giron. He married in December 1967, and he and his wife – Maria Lobaina – have two sons – Orlando, born in 1969 or 1970, and Arnaldo, born in 1968 or 1969. Tamayo-Méndez has the distinction of being the first coloured astronaut or cosmonaut to make a spaceflight.

Romanenko described Tamayo-Méndez as 'Having a great capacity for work, lives his life to the full . . . What I like about Arnaldo Tamayo is his exceptional sense of purpose. He's a very determined individual, but also very kind, and has a sharp sense of humour. Of course, as a real Cuban, he is temperamental but is in good command of his emotions. In training, Arnaldo operated with a calculating and cool mind. One of the major factors in our success was Arnaldo's single-mindedness and his amazing ability to keep calm.'

Alexei Leonov described Tamayo-Méndez as a man with determination, exceptional industriousness, and a high sense of responsibility. Other reports spoke of him as a very unassuming, affectionate and happy man, with a tremendous capacity for work, and the ability to carry out his tasks successfully.

In Spanish-speaking countries, it is the normal practice for a child to take its mother's family name as the last name. Nonetheless, it is the father's name, placed next to last, which is regarded as the surname, or family name, under which entry is made in alphabetical lists. Therefore, Arnaldo Tamayo-Méndez should be referred to as Tamayo or Tamayo-Méndez but never as Méndez

Arnaldo Tamayo-Méndez was the 7th Intercosmos cosmonaut.

Colonel-Engineer Valentina Vladimirovna Tereshkova

Valentina Tereshkova was born on 6 March 1937, in Maslennikovo, a village in the Tutayev District of the Yaroslavl Region. Her father, a tractor driver on a collective farm, was killed in action during World War 2. Her mother Yelena Fyodorovna, a weaver in a textile plant, later moved to the town of Yaroslavl, where Valentina began school in 1945. She had a younger brother, Vladimir (Volodya), and an elder sister.

Starting work at the age of 16, at the Yaroslavl tyre factory, Tereshkova continued her studies at evening school. In 1955, she transferred to the Krasnyi Perekop (Red Canal) Cotton Mill, where she operated a loom, and at the same time undertook a correspondence course at a specialist technical school. She took up parachuting at the local Aviation Club, and made her first jump on 21 May 1959. She soon became head of the Textile Mill Workers Parachute Club which she had helped to set up. By 1961, she was a graduate cotton-spinning technologist, and was also secretary of the local Komsomol (Young Communist League) committee.

It is said that following the flight of Vostok 1, Tereshkova dreamed of becoming a cosmonaut. Then, following the flight of Vostok 2, she impulsively wrote a letter to Moscow indicating her desire to train for a spaceflight. She was successful, and together with three other women, was selected to join the cosmonaut team on 16 February 1962. In March, she received a telegram telling her to report to Lieutenant-General Kamanin in Moscow. He greeted her affably, 'You're the first girl in the group so far. The rest will arrive in a couple of days.' Later the same day, she met Nikolayev, Popovich and Bykovsky for the first time. She and the other three female cosmonaut candidates were later flown to Star Town, and the first training session was held on 14 March 1962.

Tereshkova went through the complete training routine – centrifuge, isolation, weightless flights, and parachute jumps wearing a space-suit. According to one report, she sometimes found it difficult to understand rocket technology and the more complicated spacecraft systems, but she managed to overcome her difficulties by hard work.

During her training, she made over 120 parachute jumps and underwent fairly extensive pilot training. She first flew as a co-pilot on a transport plane, and then progressed to the piloting of jets. She joined the CPSU in 1962.

Her spaceflight began with the launch of Vostok 6 on 16 June 1963, in a joint flight with Vostok 5, carrying Valery Bykovsky. During this, the world's second joint flight, the two cosmonauts regularly relayed TV pictures to Earth, and conversed with each other. In addition, Tereshkova carried out a series of physiological, vestibular and psychological tests, providing data on the reactions of the female organism to spaceflight. Although suffering some disorientation and space-sickness, she was probably not so violently ill as some Western reports have suggested. She returned to Earth after a flight lasting 2 days 22 hours and 50 minutes. The mission was originally intended to last only one day, but owing to the fact that she withstood the first day well, it was agreed that the mission would be extended to three days.

Upon her return, Tereshkova was awarded the title of Hero of the Soviet Union and also received the Order of Lenin. She was also given the honorary rank of Junior-Lieutenant, and is now a Colonel-Engineer in the Air Force. After the flight, she apparently gave up her temporary role of cosmonaut, and is no longer in active training.

According to one report, (by a Soviet defector), Tereshkova was originally the back-up for Vostok 6, and replaced the prime candidate only days before launch, when she became indisposed.

In her autobiography, she later recalled how she came to be chosen for the spaceflight: 'Early 1963, I was ready. I joined a group of candidates which was in training for a space mission of rather long duration by two vehicles, one manned by a man, the other by a woman. The two female candidates were available, my friend Tanya and me. We both were fully trained in the handling of the Vostok systems. Why the Government commission assigned me I don't know. It was a last minute choice. A few female candidates were trained in the space centre and it was my sure and certain opinion that my best friend Tanya Tortchillova was the prime candidate to pilot the Vostok 6. But finally the choice was fixed upon me.'

On 3 November 1963, Tereshkova married cosmonaut Andrian Nikolayev, the pilot of Vostok 3, but in June 1983, it became known that the couple had divorced. They have a daughter, Yelena Andrianovna (known as Alyonka), a former medical student, born on 8 June 1964. It is said that Tereshkova went to great lengths to keep Yelena out of the public eye during her childhood. In June 1983, the Soviet press released the first pictures of her since she was a baby. But it apparently took three weeks to persuade Tereshkova to allow them to be taken.

There has even been speculation that the marriage of Nikolayev and Tereshkova was an arranged one, insisted on for propaganda purposes, and more importantly, for medical reasons. It was a unique chance to study the possible effects of spaceflight on the body's reproductive system, as in this case, both parties had been into space. If this were true, it might explain Tereshkova's subsequent divorce, and reluctance to expose her daughter to public gaze.

In 1964, Tereshkova enrolled at the Zhukovsky Military Air Academy, and she graduated with honours in 1969.

In the years immediately following her spaceflight, she went on a series of world tours, and in February 1964, she came to London to receive a Gold Medal from the British Interplanetary Society, commemorating her achievement of being the first woman in space. While on a visit to Havana, Cuba, she gave a speech in which she announced that the Soviet Moon team had been picked. And, she claimed, she was on it. However, any such team was subsequently disbanded when the Soviet lunar programme was cancelled.

In 1966, she became a member of the World Peace Council. She was also elected as member for Yaroslavl to the Supreme Soviet in 1967, and served on the council of the Union of the Supreme Soviet from 1966-70, and from 1970-74. She was then elected to the Presidium of the Supreme Soviet in 1974.

Since her spaceflight, she has been actively engaged as an ambassador-of-goodwill for her party and government, and has been particularly active amongst international Women's Organisations, including, since 1965, the Soviet Women's Committee. She has been chairman of the Committee since 1968. In 1975, she represented the Soviet Union at the UN Conference held in Mexico City to commemorate International Women's Year.

She is also currently a deputy to the Supreme Soviet, a member of the Presidium of the Supreme Soviet, a member of the CPSU Central Committee, and Vice-President of the Women's International Democratic Federation. In addition, she is President of the Soviet-Algerian Friendship Society.

In 1975, the International Committee for Aeronautics and Spaceflight decorated her with its 'Wind Rose' badge, and in 1976, she was awarded a degree as a Candidate of Technical Science. In April 1977, she was presented with the United Nations Gold Medal of Peace in New York, and in 1982, she was awarded a Simba Academy International award for her contribution to the International Women's Movement. She also holds two Orders of Lenin and the Joliot-Curie Gold Medal.

A slightly stocky woman, Tereshkova has grey eyes, dark hair and, it is said, an engaging smile. Her hobbies include reading, music and the theatre. At the time of Vostok 6, they were also said to have included parachute jumping, swimming, water skiing, badminton, and most of all, the trampoline.

Valentina Tereshkova's achievement of being the first woman in space will ensure her a place in History. She was the 6th Soviet cosmonaut.

Lieutenant-General Gherman Stepanovich Titov

Gherman Titov was born on 11 September 1935, in the village of Verkhneye Zhilino, Kosikha District, in the Altai, the son of a schoolteacher. His father, Stepan Pavlovich Titov, taught Russian and German, and was a lover of poetry. He had also been a taxi-driver and a mechanic. His mother was Alexandra Pavlovna Titov. Gherman studied at the Nalobikhino secondary school, and in 1949, enrolled in an elementary aviation school.

As a youth, he was involved in a bicycle accident, and broke his wrist. Rather than seek treatment, he apparently 'concealed his pain and scorned a splint or bandage.' Instead, he tried to overcome the pain and the weakness in his wrist by secret exercises. After his selection as a an aviation cadet in 1953, he was so afraid that his wrist would ruin his chances of becoming a pilot that he got up an hour early each morning to work out alone on the parallel bars in the gym. From 1955 – 57, he was a cadet at the Volgograd Air Force Pilots School, and he graduated with distinction.

He was selected to join the cosmonaut team in 1960, and his weak wrist escaped detection during the exacting physical examinations. During training, he managed to set a new record by spending 15 days in the isolation chamber.

His first assignment was as back-up to Yuri Gagarin, the pilot of Vostok 1, launched on 12 April 1961. Titov joined the CPSU shortly before his own spaceflight, which came as pilot of Vostok 2, launched on 6 August 1961. During the flight, the heating system malfunctioned, and the cabin temperature fell to 43° (6°C). This added to Titov's discomfort as he was already suffering from motion sickness, and experiencing spatial disorientation. He recovered slightly after a period of sleep, but experienced nausea throughout the mission. Nevertheless, he managed to carry out many experiments and Earth observations, before returning to Earth after a flight lasting 1 day 1 hour and 18 minutes. He is said to have suffered from inner-ear trouble for some time after his spaceflight.

Upon his return, Titov was awarded the title of Hero of the Soviet Union and also received the Order of Lenin. Soviet space doctors disclosed afterwards that if his wrist injury had been detected earlier, he would not have been allowed to fly the mission.

In 1964, he was made commander of the cosmonaut team, and very little more was heard of him. One Czechoslovakian source actually reported in 1973 that he had retired from active service, probably in 1966. He served on the Council of the Union of the Supreme Soviet from 1962 – 66, and 1966 – 70.

In August 1969, in a Tass report Titov said he was taking part in the preparations for sending more crews into outer space, 'giving his experience and knowledge to new explorers of the Universe.'

In 1971, a Novosti report stated that he was flying 'the latest Soviet supersonic jets including variable-geometry combat aircraft.' It also stated that Titov, who held the rank of test-pilot (3rd class), was studying at the Soviet General Staff Military Academy. The report led to some speculation that Titov might be preparing to lead a team of test-pilots concerned with development work on a Soviet space shuttle. He graduated from the Academy in 1972, and has a Candidate of Military Science degree. In July 1976, he served as CapCom for the Soyuz 21 / Salyut 5 mission.

Titov is currently deputy editor of the Soviet Air Force magazine 'Aviation and Cosmonautics'. He is also Chairman of the Central Board of the Soviet-Vietnamese Friendship Society. In the period 1982-83, he is known to have held an important post at the Soviet Ministry of Defence, and in 1983, in the elections to the Supreme Soviet, he was returned from an area covering Tyuratam, indicating that he was still working at the cosmodrome.

He is married, and his wife Tamara Vasilyevna was born in 1937. During his cosmonaut training, he was sent to study at the Zhukovsky Military Air Academy, and while there, his wife gave birth to their first child, a son, Igor. Unfortunately, the child became ill and died after only a few months, and Titov was temporarily taken out of the cosmonaut team. They subsequently had two more children, Tatyana, born on 23 September 1963, and Galina, born on 14 August 1965. Titov graduated from the Academy in 1968.

He is said to be a well-read and sophisticated man, who is very fond of poetry and is a Nature-lover. He is shy, although he can be hot-tempered and outspoken at times. Gagarin, upon meeting Titov was very impressed by the fact that his new acquaintance recited poems very well and knew a great many of them by heart. Not only that, but he was a remarkable athlete, as he soon demonstrated on the gymnastic apparatus. On one occasion, it is said, he performed such acrobatic feats that Gagarin, applauding excitedly, said 'Gherman, you should join the circus right now !'

He was once asked about his imya – Gherman – and explained 'My father gave me that name in honour of the character created by Pushkin.' (i.e. in Pushkin's story 'The Queen of Spades.')

Gherman Titov was the 2nd Soviet cosmonaut.

Colonel Vladimir Georgyevich Titov

Vladimir Titov was born on 1 January 1947, in the town of Sretensk, Chita Oblast. As a youth, he worked as a stoker on an oil prospecting expedition, responsible for providing heating for the drilling rig. After graduating from the Higher Air Force College, in Chernigov, in the Ukraine in 1970, he served in the Soviet Air Force as a pilot-instructor, and later, as a flight-commander with an air regiment. He has flown several types of aircraft and has the qualifications of 'Air Force pilot, 1st class' and 'Test Pilot, 3rd class.' He joined the CPSU in 1971.

He was selected to join the cosmonaut team in 1976. *It is believed that his first assignment was as a member of the support crew for Soyuz 35, but when the original back-up commander was physically disqualified, Titov moved up to become back-up commander for the mission, launched on 9 April 1980. (see section entitled 'Salyut 6 Crewing' for more details.)*

His next assignments were as the back-up to Vladimir Kovalyonok, the commander of Soyuz T-4, launched on 12 March 1981, and then as back-up to Anatoly Berezovoi, the commander of Soyuz T-5, launched on 13 May 1982.

A problem arose with Salyut 7's solar panels, resulting in a reduced flow of electrical power. This seriously affected the environmental control system, resulting in cold, damp and uncomfortable conditions which threatened to make the station uninhabitable, and also posed a threat to the continued function of the station's electronics and other systems.

To overcome this serious problem, a repair crew of Titov and Gennady Strekalov were assigned to train for a mission launched specifically to repair the faulty Salyut 7 solar array. Titov therefore made his first spaceflight as commander of Soyuz T-8, launched on 20 April 1983. Together with Strekalov and Alexander Serebrov, he was supposed to dock with Salyut 7, but once in orbit, the Soyuz rendezvous radar antenna failed to deploy properly. Several attitude control manoeuvres at high rates were made but failed to swing the boom out. (The post-flight inquiry later discovered that the whole of the antenna had been torn off when the Soyuz payload shroud separated.) The crew received permission from FCC to attempt a rendezvous using only an optical sight and ground radar inputs for guidance. However, the final approach had to be made in darkness, and Titov believed that the closing speed was too great. He therefore attempted a braking manoeuvre, but felt that the two spacecraft were still closing too fast. He aborted the rendezvous to avoid a crash, and no further attempts were made before the three men returned to Earth after a flight lasting 2 days and 18 minutes.

Upon his return, Titov did not receive any of the usual Soviet awards, indicating official dissatisfaction with the failure of the mission. He and Strekalov were then apparently 'recycled' and flew on the next available mission in a second attempt to carry out the repairs that they had been trained for.

The two men were to have been launched onboard what should have been Soyuz T-10 on 27 September 1983. However, a fire started in the base of the launch vehicle when a valve in the propellant line failed to close at T-90 seconds. This malfunction ignited a large fire which engulfed the rocket, and the automatic abort sequence failed as the wires involved burnt through. The launch controllers were forced to manually abort the mission by sending radio commands from the launch blockhouse.

The Soyuz descent module was pulled clear by the launch escape system, and the crew landed safely some 2.5 miles (4 km) from the rocket, which apparently exploded seconds after the Soyuz separated. The two men were given a medical check-up, but had sustained no injuries. Had the mission succeeded, it appears that Titov and Strekalov would have become the new resident crew, thus achieving the first handover of a Salyut from one crew to another.

In September 1985, Gennady Strekalov and Alexander Serebrov acted as commentators on Soviet TV during the Soyuz T-14 docking, and it is thought that they may have been members of the back-up crew for that mission and also the original Soyuz T-13 mission, before it was cancelled. Although very speculative, if they were the back-ups, it seems very likely that Titov would have been their commander, based on the three men's previous assignments.

Titov is married, and his wife's name is Alexandra. He is known to his friends as Volodya.

Vladimir Titov was the 54th Soviet cosmonaut, and he has spent a total of 2 days and 18 minutes in space (excluding the Soyuz launch abort.)

Colonel Pham Tuan

Pham Tuan was born on 14 February 1947 into a peasant family in the village of Quoc Tuan, Kien Xuong District, Thai Binh Province, North Vietnam, one of three children. He was said to be an excellent pupil at high school and was admitted into the Vietnamese Communist Youth Union in 1963. In the summer of 1965, he was enrolled in the Vietnamese Air Force, but was disqualified after his initial medical revealed a faster than normal heartbeat. Tuan asked for a private meeting with the physician, where he argued his general fitness and good health. The doctor then re-examined him and found that Tuan's rapid heartbeat was simply due to his over-excitement ! He was passed as fit for military service.

This delay caused him to arrive at the Soviet combat pilots school three months late, when his colleagues had already completed their theoretical training. However, in all the theoretical subjects Tuan reportedly got the highest marks. He did well in both day and night flying and graduated in May 1968 with a pilots certificate which stated that he was a 'strong-willed and very promising' pilot.

Following graduation, Tuan returned to Vietnam and was promoted to Sub-Lieutenant, and was assigned as a fighter-pilot in the 'Red Star' regiment – a hunter-interceptor unit of the Vietnamese Air Force. He joined the Vietnamese Communist Party in November 1968, and was also promoted to Lieutenant in the same month. He flew as wing man for Vu Ngoc Dinh, one of the North Vietnamese Air Force's best pilots, and during an air battle on 28 January 1970, Tuan claimed a US F4 Phantom, although this is not confirmed by US records.

In late 1971, he was assigned to a unit of the 'Red Star' regiment specialising in night flights. In December 1972, during the Vietnam War, Tuan returned to base after a sortie to find the runway bombed. He landed, but hit a crater, causing his plane to flip over and skid for over 1000ft (305 m) before rolling upright. Luckily, he escaped unhurt. A week later, during the night of 27 December, it is claimed that Tuan became the first and only man in the world to shoot down a USAF B-52 bomber. The USAF however strongly deny this claim, saying that all 17 B-52's lost to enemy fire were downed by surface to air missiles, and not by hostile aircraft.

For this feat, Tuan was awarded the title of 'Hero of the Socialist Republic of Vietnam', the 'Order of Military Exploit, 3rd Class' and the 'Order of Glorious Combatant' 1st, 2nd and 3rd Class. He was also promoted to Lieutenant-Captain, and then Captain in 1974. In 1976, he was appointed deputy-commander of the 'Red Star' regiment, and was then promoted to Major. In 1977, he was sent for training at the Gagarin Air Force Academy in the Soviet Union, and was promoted to Lieutenant-Colonel in 1979.

In May 1979, Vietnam became a member of the Intercosmos programme, and as such was immediately invited by the Soviet Union to participate in a joint spaceflight, in the same way as the other member countries of Intercosmos. It was then revealed that Vietnamese cosmonauts were already undergoing training at the Yuri Gagarin Cosmonaut Training Centre.

Pham Tuan and Bui Thanh Liem were the two cosmonauts, and they had begun their training in April 1979. Tuan was paired with Viktor Gorbatko in September 1979, and eventually flew as research-cosmonaut onboard Soyuz 37, launched on 23 July 1980. He and Gorbatko linked-up with the orbiting Salyut 6 space-station and joined the crew of Popov and Ryumin already on board. They then carried out a series of joint Soviet-Vietnamese experiments and observations before returning to Earth after a flight lasting 7 days 20 hours and 42 minutes.

Upon his return, Tuan was awarded the title of Hero of the Soviet Union and also received the Order of Lenin. He was also awarded the titles of Hero of the Socialist Republic of Vietnam (for the second time), and Hero of the Vietnamese People's Armed Forces. He was also awarded the Ho Chi Minh order, 1st class, and was made an honorary citizen of Dzhezhazgan.

He is married, and he and his wife – a Lieutenant in the Army Medical Corps – have a daughter, born in 1975 or 1976. Tuan has been described by Gorbatko as 'an able, intelligent, modest and industrious pilot.....diligent and persistent.'

Pham Tuan was the 6th Intercosmos cosmonaut.

Lieutenant-Colonel Vladimir Vladimirovich Vasyutin

Vladimir Vasyutin was born on 8 March 1952, in Kharkov, in the Ukraine.

He joined the Soviet Air Force, and in 1969, enrolled at the Kharkov Higher Air Force Pilots School. After graduating, he served in the Air Force as a pilot-instructor. He is said to have mastered several types of aircraft and has the qualifications of Military Pilot 1st Class and Test-Pilot 3rd Class. He joined the CPSU in 1972.

He was still serving as a pilot-instructor when he was selected to join the cosmonaut team in 1976, and he reported for duty in February of that year. He also enrolled at a Test-Pilots school in 1976, and graduated in 1977.

Vasyutin's first assignment was as the back-up to Leonid Popov, the commander of Soyuz T-7, launched on 19 August 1982. The original back-up was Yuri Romanenko, but it appears that he was physically disqualified shortly before the flight, and was replaced by Vasyutin *who may originally have been the support crew commander.*

It is thought that Vasyutin may also have served as commander of the support crew for Soyuz T-9, launched on 27 June 1983. He should then have been assigned to the back-up crew for Soyuz T-10, but following the 're-cycling' of Titov and Strekalov, he is believed to have been given a second support crew assignment on that mission instead. (see section entitled 'Salyut 7 Crewing' for more details.) However, when the mission suffered a launch abort, he was given the back-up assignment for the replacement Soyuz T-10, launched on 8 February 1984.

He then served as the back-up to Vladimir Dzhanibekov, the commander of Soyuz T-12, launched on 17 July 1984.

When the crewing tables are examined, it appears that Vasyutin was in training for a long-duration mission, and was temporarily sidetracked by the need to fly the Soyuz T-12 mission. Upon the completion of that flight, the original schedule was resumed, and he should then have flown on Soyuz T-13 with Viktor Savinykh and Alexander Volkov. However, when Salyut 7 suffered an onboard failure which left it without power, a repair mission became necessary and the crew were split up. Savinykh flew the replacement Soyuz T-13 mission with Vladimir Dzhanibekov, and it is believed that Vasyutin was paired with Georgi Grechko on the back-up crew. Volkov was stood down. (See section entitled 'Salyut 7 Crewing' for more details.)

Once the Salyut had been repaired, the Soviets wanted to get back to the original flight schedule, and Vasyutin's next assignment therefore came as commander of Soyuz T-14, launched on 17 September 1985. At the time, it was reported that he had been in training for the mission since September 1984. Together with Volkov and Grechko, he linked up with the orbiting Salyut 7 space-station and joined the crew of Dzhanibekov and Savinykh already onboard. The two crews worked together for a week, and then Grechko returned to Earth with Dzhanibekov, the Soyuz T-13 commander, thus carrying out the first partial replacement of a crew. Vasyutin and Volkov remained onboard the Salyut with Viktor Savinykh, thus restoring the original Soyuz T-13 crew.

The three men settled down to a long programme of experiments and observations, and this included working with the Cosmos 1686 space-station module.

From later Soviet reports, it is clear that the crew fully expected to still be in orbit on 1 January 1986, and it seems likely that Savinykh was intended to set a new duration record. However, in November, Vasyutin became ill and the mission had to be terminated early. The crew began scrambling their conversations with FCC several days before returning to Earth, Vasyutin having spent 64 days 21 hours and 52 minutes in space.

It was the first time in either the Soviet or American programmes that a spaceflight had had to be cut short due to illness, and the nature of Vasyutin's complaint was not at first revealed. Following a preliminary examination at the landing site, doctors would only say 'Vladimir Vasyutin's condition is satisfactory . . . we cannot yet say anything more than that as we have to make a thorough check-up. It is only clear now that, as we expected, he needs hospital treatment.'

A later report stated that he was suffering from some internal inflammation, and that his condition would require a thorough examination and treatment. This was at first understood to indicate appendicitis.

However, the true nature of his illness was made public some six weeks after the return to Earth. Savinykh kept a diary onboard Salyut 7, and extracts were published in *Pravda* . He recalled why the mission was terminated early: 'It all began when we observed a slight uneasiness in Volodya Vasyutin's behaviour. There was a loss of sleep and appetite. We thought it was all to do with his mood – it happens to us all. We tried to buck him up by cracking jokes and giving bits of advice. But he became ill. Volodya wanted to see it through but it became more difficult for him.'

Savinykh went on to say that Vasyutin became 'a bundle of nerves,' and it was hard to look at him. After consulting with FCC, it was recommended that Vasyutin should be brought back for treatment.

His condition gave more cause for concern, and Savinykh made the following diary notes on 18 November: 'I couldn't write yesterday and it's the same today. Yesterday we decided to come down. We gave ourselves three days to prepare...During that time it seemed to us, the doctors and to Volodya himself at times that he was improving. But still we made the right decision.'

In light of the psychological problems that he obviously experienced onboard Salyut 7, it seems most unlikely that Vasyutin will be allowed to make another spaceflight.

He was later awarded the title of Hero of the Soviet Union and also received the Order of

Lenin. He is married, and his wife works at the Hydrotank at the Yuri Gagarin Cosmonaut Training Centre. They have two children.

Vladimir Vasyutin was the 59th Soviet cosmonaut.

Igor Petrovich Volk

Igor Volk was born on 12 April 1937, in the town of Gottwald in the Kharkov Region of the Ukraine. He spent his childhood in the Far East, where his family lived for 14 years.

It is said that it was by sheer chance that he began his life-long interest in flying. A flying club instructor visited his school, following which Volk went to the flying club and fell in love with aviation. He joined the Kursk flying club, making his first flight in April 1954, and went on to the Kirovograd Military Flying School from which he graduated in 1956. He then served in the Soviet Air Force for seven years, before joining a Test-Pilots School in 1963. At the same time, he enrolled for evening classes at the Moscow Aviation Institute. He joined the CPSU in 1964, and qualified as a test-pilot in 1965.

He graduated from the Moscow Aviation Institute in 1969, and received a degree as a Candidate of Science (Engineering). He then served as a test-pilot for many years, until he was selected to join the cosmonaut team in December 1978. Before his selection, he decided to fly with a group of cosmonauts onboard a laboratory aircraft to find out what weightlessness was like. He decided at the time that it was not for him, it was for 'daredevils !' He later relented however.

His first spaceflight came as research-cosmonaut onboard Soyuz T-12, launched on 17 July 1984. Volk, Vladimir Dzhanibekov and Svetlana Savitskaya linked up with the Salyut 7 space-station and joined the crew of Kizim, Solovyov and Atkov already onboard.

It is possible that Dzhanibekov and Volk were also the support crew on Soyuz T-10, launched on 8 February 1984, and were training for a long-duration flight. They were then apparently sidetracked by the need to fly a repair mission following a main fuel line failure on the Salyut. (see section entitled 'Salyut 7 Crewing' for more details.)

Dzhanibekov and Savitskaya carried out an EVA during which they tested a new electron beam device, using it to carry out a series of operations involving the cutting, welding and soldering of metal plates and the spraying on of a coating. In addition, the whole crew, including Volk, conducted a series of experiments onboard the Salyut before returning to Earth after a flight lasting 11 days 19 hours and 14 minutes.

Upon his return, Volk was awarded the title of Hero of the Soviet Union and also received the Order of Lenin.

He had first become acquainted with Svetlana Savitskaya some years earlier, when they worked together as test-pilots. 'I met him earlier than anyone else in the cosmonauts detachment,' she said. 'When I was training at the Test Pilots School, he gave the graduation examination to our course.'

Dzhanibekov has described Volk as 'an efficient and responsible researcher,' whereas Vladimir Solovyov reported that 'Igor is a very athletic person, he can't understand how anyone can begin life in the morning without exercises.' He is said to be of average height, solidly built, with soft features and reddish hair. He is married and has a child.

Volk holds the qualification of 'Test-Pilot 1st Class' and the honorary title of 'Merited Test-Pilot of the Soviet Union.' He has logged over 4700 hours flying time, 2700 of these in test-flights, and is said to be a highly skilled test-pilot, held in high esteem by his colleagues.

He has apparently flown nearly all types of Soviet aircraft, and during one flight, he was ordered to eject from a supersonic aircraft when the tyres flew off during take-off. He refused, and managed to land the plane on bare wheel-drums.

According to one Western report, he is thought to have worked on the Soviet TU-144 aircraft.

Of his profession as a test-pilot, Volk once said 'It is just a job, like any other. Critical situations do arise, but they are not as frequent or as dramatic as is generally believed. Not everyone could do the work I admit. You have to think fast, but the decisions you make must always be correct. When flying you are always working against time.'

Igor Volk was the 58th Soviet cosmonaut.

Lieutenant-Colonel Alexander Alexandrovich Volkov

Alexander Volkov was born on 27 May 1948, in the mining town of Gorlovka, in the Ukraine.

After graduating from the Kharkov Higher Air Force Pilots School in 1970, he served in the Air Force as a pilot-instructor at the School. He is said to have mastered several types of aircraft and has the qualifications of Military Pilot 1st Class and Test-Pilot 2nd Class. He joined the CPSU in 1973.

He was still serving as a pilot-instructor when he was selected to join the cosmonaut team in 1976, and he reported for duty in February of that year.

Details of cosmonauts are rarely obtained prior to them making a first spaceflight. However, at one stage, several Western observers identified the Soyuz T-8 back-up crew, and named Volkov as the back-up research-cosmonaut. Few other details about him were available until the launch of Soyuz T-14. At that time however, it was announced that Volkov had been training for Soyuz T-14 for a year – since September 1984 – and that it was his first training programme i.e. he had **not** been in any previous back-up crews. *(It now seems likely that Viktor Savinykh was the back-up research-cosmonaut on Soyuz T-8.)*

Volkov, Vladimir Vasyutin and Viktor Savinykh were the prime crew for the original Soyuz T-13 mission. However, when Salyut 7 suffered an onboard failure which left it without power, a repair mission became necessary and they were split up. Savinykh flew the replacement Soyuz T-13 mission with Vladimir Dzhanibekov, and it is believed that Vasyutin was paired with Georgi Grechko on the back-up crew. Volkov was stood down. *(See section entitled 'Salyut 7 Crewing' for more details.)*

Once the Salyut had been repaired, the Soviets wanted to get back to the original flight schedule, and Volkov's next assignment therefore came as research-cosmonaut onboard Soyuz T-14, launched on 17 September 1985. Together with Vasyutin and Grechko, he linked up with the orbiting Salyut 7 space-station and joined the crew of Dzhanibekov and Savinykh already onboard. The two crews worked together for a week, and then Grechko returned to Earth with Dzhanibekov, the Soyuz T-13 commander, thus carrying out the first partial replacement of a crew. Vasyutin and Volkov remained onboard the Salyut with Viktor Savinykh, thus restoring the original Soyuz T-13 crew.

The three men settled down to a long programme of experiments and observations, and this included working with the Cosmos 1686 space-station module.

From later Soviet reports, it is clear that the crew fully expected to still be in orbit on 1 January 1986, and it seems likely that Savinykh was intended to set a new duration record. However, in November, Vladimir Vasyutin became ill and the mission had to be terminated early. The three men returned to Earth, Volkov having spent 64 days 21 hours and 52 minutes in space.

Upon his return, he was awarded the title of Hero of the Soviet Union and also received the Order of Lenin.

Volkov is married, and he and his wife, Anna, have two sons, Dima and Sergei.

Alexander Volkov was the 60th Soviet cosmonaut.

Vladislav Nikolayevich Volkov (1935-1971)

Vladislav Volkov was born on 23 November 1935, in Moscow. His father Nikolai was an aeronautical engineer, and his mother Olga worked at aircraft plants for many years. The family lived within earshot of Tushino Airport just outside Moscow, and Volkov's first word, his father later joked, was not 'Mama' but 'Samalot' (Airplane). He apparently saw his first jets in August 1946 when he was 10. All of his neighbours were aeronautical engineers or pilots, and this heavily influenced his choice of career.

In 1953, he graduated from Moscow High School No.212 and had to choose between becoming an aviation cadet in the Air Force or studying for an engineering degree first. Upon the advice of his parents and teachers, he enrolled at the Moscow Aviation Institute in 1955. While there, he joined an aeroclub and earned a propeller aircraft certificate, and also qualified as a parachutist. He was also interested in soccer and hockey and his skilful performance on the ice and the field was soon reported by the press. According to some, he could easily have made a name for himself as an athlete.

During his studies at the Institute, he met and married another engineering student, and he and Lyudmilla both graduated in 1958, the same year their son Vladimir was born. Following graduation, Volkov began work at the Korolyov design bureau in Moscow, where he was involved in aviation design work. In 1958, the bureau was given the task of building the command module for a manned spaceflight, and for the next few years Volkov worked on the construction and testing of the Vostok and Voskhod spacecraft. In 1962, he went to the cosmodrome to watch the launches of Vostok 3 and 4.

He was closely involved in the development of the Voskhod spacecraft, sitting in the flight simulators as often as possible, and according to one report, he and Feoktistov both volunteered to fly in Voskhod 1. Both men were interviewed in 1964, but Feoktistov was chosen for the mission, and Volkov wasn't officially selected to join the cosmonaut team until 1966. He joined the CPSU in 1965.

He reported to Star Town on 5 September 1966, and when asked about his decision to become a cosmonaut, he said 'I want to master space equipment, to test it and help in improving it, to do something resembling what test-pilots now do in aircraft design offices.'

His first spaceflight came as flight-engineer onboard Soyuz 7, launched on 12 October 1969. Volkov, Anatoli Filipchenko and Viktor Gorbatko carried out rendezvous manoeuvres and other experiments in a joint flight with the Soyuz 6 and 8 spacecraft, before returning to Earth after a flight lasting 4 days 22 hours and 40 minutes.

Upon his return, Volkov was awarded the title of Hero of the Soviet Union and also received the Order of Lenin. His next assignment was as the back-up to Alexei Yeliseyev, the flight-engineer onboard Soyuz 10, launched on 23 April 1971.

His second spaceflight came as flight-engineer onboard Soyuz 11, launched on 6 June 1971. *It is believed that he was originally the back-up flight-engineer for both Soyuz 10 and Soyuz 11, and was not actually scheduled to fly to Salyut 1. (see section entitled 'Salyut 1 Crewing' for more details.)*

The Soyuz 11 crew docked with the Salyut 1 space-station, and spent nearly 23 days onboard, carrying out Earth observations work and other experiments. Then Volkov, Georgi Dobrovolsky and Viktor Patsayev reboarded their Soyuz spacecraft and returned to Earth. Tragically, a pressure equalization valve opened prematurely while the descent module was still above the atmosphere, and the crew were killed following the loss of cabin oxygen and pressure. The death of the three men on 30 June 1971 was a severe blow to the Soviet manned space programme, and delayed further flights for over two years. The Soyuz 11 mission had lasted 23 days 18 hours and 22 minutes, a record which stood for nearly two years before being broken by the American Skylab 1 astronauts in 1973.

Volkov was posthumously awarded the title of Hero of the Soviet Union and also received the Order of Lenin, both for the second time. Following a military state funeral, his ashes were buried in the Kremlin Wall. In his honour, the Soviets named one of the space tracking vessels the 'Cosmonaut Vladislav Volkov', and a street in Kaluga, the birthplace of Tsiolkovsky, was also named after him.

He was married, and he and his wife Lyudmilla Alexandrovna – an engineer – had a son, Vladimir, born in 1958.

Volkov was said to be an athletically-built, outgoing and fun-loving man, with a good sense of humour. This, together with his willingness to help, made him a very popular man on the cosmonaut team. A guitarist, he often played and sang for his family and friends and reporters at the space centre.

The surname 'Volkov' means 'Wolf' in Russian, and the imya 'Vladislav' means 'Prince of Slavs'. Volkov was not keen on this and tried out various nicknames and variations, before adopting his own choice of imya – Vadim – by which his friends knew him. It was derived from a French name, the source of which he later could not recall.

His autobiography entitled 'We're Stepping into the Sky' was published posthumously, having only been completed a few days before the launch of Soyuz 11.

Vladislav Volkov was the 20th Soviet cosmonaut, and he spent a total of 28 days 17 hours and 02 minutes in space.

Colonel Boris Valentinovich Volynov

Boris Volynov was born on 18 December 1934, in Irkutsk, Siberia, and was brought up by his mother Yevgenia alone. She was trained as a pediatrician, but as the small town needed a traumatologist, she learned the new profession, and during World War 2, worked as a surgeon. She is now retired. After a ten-year education in Prokopyevsk, Volynov applied to the regional Komsomol committee for a letter of recommendation to a fighter pilot school, and this was granted. He attended the Volgograd Air Force School, and following graduation, he served for four years as a fighter-pilot. He joined the CPSU in 1958.

He was selected to join the cosmonaut team in 1960, and was the second man appointed, Popovich being the first. His first assignment was as the support cosmonaut for Vostok 4, launched on 12 August 1962. He then served as back-up to Valery Bykovsky, the pilot of Vostok 5, launched on 14 June 1963, and then as back-up to Vladimir Komarov, the commander of Voskhod 1, launched on 12 October 1964.

It has been revealed that a *third* mission in the Voskhod series was planned for the early summer of 1966, involving a two-man team of cosmonauts in a long-duration mission. A training group was formed, probably around April 1965, but the mission was cancelled in early 1966.

The four cosmonauts since named as being part of this group were Volynov, Shonin, Beregovoi and Shatalov. It is thought that the prime crew would have been Volynov and Shonin, with Beregovoi and Shatalov as the back-ups, as Volynov had already served as back-up for the Vostok 5 and Voskhod 1 missions. Following cancellation of the remaining Voskhod missions, the four men were re-assigned to the new Soyuz programme.

Volynov's next assignment was as support cosmonaut for Soyuz 3, launched on 26 October 1968. He also graduated from the Zhukovsky Military Air Academy in 1968.

His first spaceflight came as commander of Soyuz 5, launched on 15 January 1969. The spacecraft docked with Soyuz 4, carrying Vladimir Shatalov, in the world's first docking of two manned spacecraft. Yevgeny Khrunov and Alexei Yeliseyev then carried out an EVA to transfer from Soyuz 5 to Soyuz 4, and joined Shatalov inside. The two craft then separated, and Volynov, alone in Soyuz 5, later returned to Earth after a flight lasting 3 days and 54 minutes. Upon his return, he was awarded the title of Hero of the Soviet Union and also received the Order of Lenin.

Following his flight, Volynov was put in charge of the 1970 cosmonaut selection, and in particular, he took part in preparing Dzhanibekov and Romanenko for their back-up assignments during the Apollo-Soyuz Test Project (ASTP). He also served on the ground control teams for several Soyuz missions.

It is thought that Volynov may have trained for Salyut 2, but when that failed, he was re-assigned to Salyut 3. (see section entitled 'Salyut 2/Cosmos 557 Crewing' for more details.)

His next assignment was as back-up to Pavel Popovich, the commander of Soyuz 14, launched on 3 July 1974, and he also served as CapCom for that mission. *It is believed that he also served as the back-up to Gennady Sarafanov, the commander of Soyuz 15, launched on 26 August 1974. (see section entitled 'Salyut 3 Crewing' for more details.)*

His second spaceflight came as commander of Soyuz 21, launched on 6 July 1976. He and Vitaly Zholobov docked with the Salyut 5 space-station, and carried out a programme of medical, biological and technical experiments and Earth resources work. Western observers believe that they also carried out military reconnaissance. The two men returned to Earth after a flight lasting 49 days 6 hours and 23 minutes.

According to one Western report, medical records for Soyuz 21 published two years after the flight show that during their third week in space, the two men were working for up to 16 hours a day (twice the recommended amount) and were foregoing their daily exercise routine. By 30th July, Volynov was actually ill, and Zholobov was 'nagged by continuing problems.'

The report suggested that the explanation for the vastly increased workload was that two Cosmos spy satellites had malfunctioned, and the cosmonauts had been ordered to fill in.

They were under severe strain, and eventually were given a week to recover. However, they were soon put under pressure again, with as little as four hours sleep a day. Both showed renewed signs of stress, and shortly afterwards were given just a few hours notice before returning to Earth. They needed over a week to fully regain their strength.

There were many indications that the return to Earth was an unplanned one, bringing a sudden end to the mission. A second Western report claimed that the crew had been forced to return under emergency conditions because of an 'acrid odour' coming from the Salyut's environmental control system. The two men tolerated the odour for some time, but were unable to find the cause of the problem before the odour became unbearable, forcing them to evacuate the Salyut.

Volynov was subsequently awarded the title of Hero of the Soviet Union and also received the Order of Lenin, both for the second time. He revealed that if he was not assigned to a further mission, he would take up science, as he had wished to do for some time, or would work as a flight control specialist.

In March 1978, he was in charge of the search headquarters for the Soyuz 28 mission, and he met the crew at the recovery site and then accompanied them back to Leninsk. He helped train the two Mongolian cosmonauts during their survival and water recovery courses, and in 1980, he was described as 'one of the top executives for training cosmonauts.' He is presently Commander of the cosmonauts detachment at the Yuri Gagarin Cosmonaut Training Centre.

He is married, and he and his wife – Tamara, a metallurgical engineer – have a son Andrei, born in 1957 or 1958, and a daughter Tatyana, born in 1964 or 1965. Volynov devotes much of his time to sports and to public affairs, and is Chairman of the Soviet Tennis Federation, and Vice-President of the Soviet-Polish Friendship Society. By making the Soyuz 5 flight, he became the first Soviet Jew to fly in space.

Boris Volynov was the 14th Soviet cosmonaut, and he has spent a total of 52 days 7 hours and 17 minutes in space.

Medical-Lieutenant Boris Borisovich Yegorov

Boris Yegorov was born on 26 November 1937, in Moscow, the son of a prominent Soviet brain surgeon. At the age of 18, he went to the Moscow First Medical Institute to become an aviation-medicine specialist, with a particular interest in the vestibular apparatus of the human ear. He graduated with a degree in aviation and space medicine in 1961.

He joined the Soviet space programme as an aviation-medicine specialist, and he was one of the doctors who examined Gagarin after his spaceflight.

In an interview in a book published in 1973, he revealed that he was originally shortlisted for a Vostok mission which was apparently cancelled. Had he flown, it is likely that the mission – probably Vostok 7 – would have been an eight day flight launched in late 1963 or early 1964. He was given a second chance with the first Voskhod mission.

'At last I was asked to join the detachment of cosmonauts' he said. 'It seemed then, that, after six months, I would definitely fly. But it was not to be. I continued my work, participating in the preparation of pilots for the flights. But then my chance came. A physician-research worker was needed in the crew. And I was one of the candidates.'

He was therefore officially selected to join the cosmonaut team in 1964 to fly as physician onboard Voskhod 1, launched on 12 October 1964. Yegorov was the first doctor to make a spaceflight, and during the mission, he studied the reactions of Vladimir Komarov and Konstantin Feoktistov. He also carried out many experiments in connection with the balancing organism in the inner ear, a subject in which he is passionately interested. The three men returned to Earth after a flight lasting 1 day and 17 minutes.

Upon his return, Yegorov was awarded the title of Hero of the Soviet Union and also received the Order of Lenin. It then seems that he gave up his temporary role of cosmonaut, and since his spaceflight, he has remained with the team of space doctors which examines each successive crew as they return to Earth. During his spaceflight training, he had made 11 parachute jumps, and was also an experienced mountaineer.

There is some confusion over Yegorov's selection date. Officially, he joined the cosmonaut team in 1964. However, he says he was selected, and then after six months, was given a Vostok assignment, subsequently cancelled. If he was selected in say, January

1964, then he would have been taken off Vostok training in July at the very earliest, assuming the Vostok mission was scrapped almost immediately after he got the assignment. Assuming an immediate *switch to the Voskhod 1 mission – which is not implied in his version of events – would only allow three months training. Yegorov does refer to a 'shortened programme of training' for the mission, but only three months seems to be unlikely in the extreme. The conclusion to be drawn is that Yegorov was originally selected earlier than 1964 – probably in 1963. (N.B. According to one source, the Voskhod training began in June 1964.)*

He is married, and he and his wife Eleanora Victorovna (or Valentinovna), a doctor in a leading Moscow eye clinic, have a son Boris, born in 1962. Yegorov received a Doctor of Medicine degree from the Humboldt University of Berlin in 1965, and in 1967, he earned a Candidate of Medical Sciences degree. He joined the CPSU in 1975, and is currently head of a laboratory.

Boris Yegorov was the 9th Soviet cosmonaut

Alexei Stanislavovich Yeliseyev

Alexei Yeliseyev was born on 13 July 1934 in the town of Zhizdra, in the Kaluga Region. As his parents both went out to work, he had to look after his younger brother, run the household, and handle the family finances.

When he was in his 8th grade, he decided to devote himself to the exact sciences. His parents encouraged him to specialise in maths and physics, and he took up sports and chess as recreations. Before deciding which institute to enter, he carefully checked what each one had to offer, and finally settled on the E.N. Bauman Moscow Higher Technical School, which he entered after finishing secondary school in 1951.

Yeliseyev was a very proficient fencer, and became the national fencing champion twice. However, he had to give up the sport when the demands of his studies on his time became too great. In 1963, he began to specialise in space-engineering, and in 1967, he earned a degree as a Candidate of Science. He also joined the CPSU in 1967.

He was interviewed with Konstantin Feoktistov in 1964, but was not officially selected to join the cosmonaut team until 1966. He reported to Star Town on 5 September 1966, and soon received his first assignment. On 23 April 1967, Vladimir Komarov was launched onboard Soyuz 1. He was brought back to Earth after just over a day in orbit, and was killed when the spacecraft's parachute tangled following re-entry. It is now clear, despite official Soviet denials at the time, that a second spacecraft, Soyuz 2, was to have been launched on 24 April. Following docking, an exchange of crews would have taken place.

The Soyuz 2 prime and back-up crews have since been identified by Western observers. The prime crew were Valery Bykovsky, Alexei Yeliseyev and Yevgeny Khrunov, and the back-up crew were Pavel Belyayev/Andrian Nikolayev, Valery Kubasov and Viktor Gorbatko. Yeliseyev's first assignment therefore was as flight-engineer for Soyuz 2, and he would have made an EVA to transfer from Soyuz 2 to Soyuz 1. However, following the initial problems with Soyuz 1, the Soyuz 2 launch was cancelled.

An unmanned Soyuz 2 subsequently flew in a rendezvous operation with Soyuz 3, piloted by Georgi Beregovoi, and in January 1969, nearly two years late, the mission intended for Soyuz 1/2 was carried out by the Soyuz 4/5 spacecraft. The prime and back-up crews of Soyuz 2 were given the same assignments on Soyuz 5 – only the commanders had changed.

Yeliseyev's first spaceflight therefore came as flight-engineer onboard Soyuz 5, launched on 15 January 1969. Soyuz 5 docked with Soyuz 4, carrying Vladimir Shatalov, in the world's first docking of two manned spacecraft. Yeliseyev and Yevgeny Khrunov then carried out an EVA to transfer from Soyuz 5 to Soyuz 4, and joined Shatalov inside. The EVA lasted 1 hour, of which 15 minutes was spent in open space. The two craft then separated, and the next day Soyuz 4 re-entered, Yeliseyev having spent 1 day 23 hours and 39 minutes in space. Upon his return, he was awarded the title of Hero of the Soviet Union and also received the Order of Lenin.

He only had to wait another nine months for his second spaceflight which came as flight-engineer onboard Soyuz 8, launched on 13 October 1969. Yeliseyev and Vladimir Shatalov carried out rendezvous manoeuvres and other experiments in a joint flight with the Soyuz 6 and 7 spacecraft, before returning to Earth after a flight lasting 4 days 22 hours and 51 minutes.

Yeliseyev was subsequently awarded the title of Hero of the Soviet Union and also received the Order of Lenin, both for the second time.

His third spaceflight came as flight-engineer onboard Soyuz 10, launched on 23 April 1971. *It is possible that Yeliseyev was originally intended to fly on the next mission, Soyuz 11, but the crew was switched due to the illness of the original Soyuz 10 crew. If this is correct, Yeliseyev may have then served as the back-up flight-engineer on Soyuz 11. (See section entitled 'Salyut 1 Crewing' for more details.)*

The Soyuz 10 spacecraft docked with the Salyut 1 space-station – the first time such an operation had taken place – but undocked after only 5 hours and 30 minutes, without the crew entering the station. The Soviets have stated that it was never their intention for the station to be manned during this initial step, but in reality it would seem likely that there was actually some problem with the transfer hatch, preventing the crew from entering the Salyut. Yeliseyev, Vladimir Shatalov and Nikolai Rukavishnikov returned to Earth after a flight lasting 1 day 23 hours and 46 minutes.

Yeliseyev was subsequently awarded a third Order of Lenin.

In the period 1973-84, he served as Flight Director during every Soviet manned spaceflight. The only exception was that during the ASTP mission, he temporarily handed control of the Salyut 4 flight to Konstantin Feoktistov.

In 1972, he earned a DSc. degree, and in 1973, he was appointed Soviet Flight Director for ASTP. In connection with that project, he served as Flight Director for the Soyuz 16 ASTP rehearsal flight in December 1974, and was responsible for the training of the Soviet FCC personnel for ASTP, and for solving the problems of joint mission control.

He also served as Flight Director for the joint Soviet-Bulgarian Soyuz 33 mission in 1979, following which he was made an honorary citizen of Lovech, Bulgaria. He served in the same capacity for the joint Soviet-Hungarian Soyuz 36 mission in 1980, and was awarded the Golden Wreath Badge of the Soviet-Hungarian Friendship Society.

It is understood that in 1984, he took up a position as a professor at the E.N. Bauman Moscow Higher Technical School, and in January 1986, he was appointed Rector.

He is married, and his wife, Larisa Ivanovna Komarovna, is a Master of Technical Sciences and works as a researcher on management problems. They have a daughter, Yelena, born in 1960.

Yeliseyev, who is said to be a man of few words, likes painting and still enjoys fencing. He is also a member of the Komsomol Central Committee. His mother, Valentina, has a DSc. degree in Chemistry, and was the chief professor in charge of a laboratory in the Physical Chemistry Institute of the Soviet Academy of Sciences.

Larisa Yeliseyeva was once asked what she thought of her husband's profession. 'When Alexei told me he was going to take a test for the cosmonaut group, I wasn't overjoyed at the news. I clearly understood what was in store for me. But at the same time I know that if a person is working on space technology, a spaceflight would undoubtedly be the crowning achievement, provided one's health is good of course. All the reasoning in the world didn't help me, though, when Alexei made his flight. I took all three flights very hard. The most difficult part is when your husband leaves for the cosmodrome. It's probably a bit easier after the spacecraft is launched because thousands of other people follow the flight. I know that they, too, are concerned about Alexei, and that helps.'

Yuri Gagarin once said of Yeliseyev 'He brings with him vast engineering and design experience which helps the crew to master the operation of the new spaceship and its systems. Alexei is a friendly person, but very demanding when it comes to work. He has become inseparable from our group. He is a valuable specialist in our space programme and we are always glad when such people join us.'

Alexei Yeliseyev was the 15th Soviet cosmonaut, and he has spent a total of 8 days 22 hours and 16 minutes in space.

Colonel-Engineer Vitaly Mikhailovich Zholobov

Vitaly Zholobov was born on 18 June 1937, in the village of Zburyevka, in the Kherson region of the Ukraine, and was brought up in Azerbaijan. His father was a sea-captain. After finishing school, he went to the Azerbaijan Oil and Chemistry Institute, and graduated in 1959 with a degree in chemistry. He then became an officer in the Soviet Army, working as a test-engineer in various military units.

He was selected to join the cosmonaut team in 1963, reporting for duty on 11 January, and he joined the CPSU in 1966. *It is thought that he may have trained for Salyut 2, but when that failed, he was re-assigned to Salyut 3. (see section entitled 'Salyut 2/Cosmos 557 Crewing' for more details.)*

It is believed that Zholobov's first assignment was as the back-up to Yuri Artyukhin, the flight-engineer onboard Soyuz 14, launched on 3 July 1974. He is then believed to have served as the back-up to Lev Demin, the flight-engineer onboard Soyuz 15, launched on 26 August 1974. (see section entitled 'Salyut 3 Crewing' for more details.)

He graduated from the V.I. Lenin Military-Political Academy in 1974, and made his first spaceflight as flight-engineer onboard Soyuz 21, launched on 6 July 1976. He and Boris Volynov docked with the Salyut 5 space-station, and carried out a programme of medical, biological and technical experiments and Earth resources work. Western observers believe that they also carried out military reconnaissance. The two men returned to Earth after a flight lasting 49 days 6 hours and 23 minutes.

According to one Western report, medical records for Soyuz 21 published two years after the flight show that during their third week in space, the two men were working for up to 16 hours a day (twice the recommended amount) and were foregoing their daily exercise routine. By 30th July, Volynov was actually ill, and Zholobov was 'nagged by continuing problems.'

The report suggested that the explanation for the vastly increased workload was that two Cosmos spy satellites had malfunctioned, and the cosmonauts had been ordered to fill in.

They were under severe strain, and eventually were given a week to recover. However, they were soon put under pressure again, with as little as four hours sleep a day. Both showed renewed signs of stress, and shortly afterwards were given just a few hours notice before returning to Earth. They needed over a week to fully regain their strength.

There were many indications that the return to Earth was an unplanned one, bringing a sudden end to the mission. A second Western report claimed that the crew had been forced to return under emergency conditions because of an 'acrid odour' coming from the Salyut's environmental control system. The two men tolerated the odour for some time, but were unable to find the cause of the problem before the odour became unbearable, forcing them to evacuate the Salyut.

Upon his return, Zholobov was awarded the title of Hero of the Soviet Union and also received the Order of Lenin. He is married, and his wife Liliya is an engineer. They married while they were both still students, and they have a daughter, Yelena. Zholobov has the unusual distinction of being the first Soviet cosmonaut to go into orbit wearing a moustache !

Vitaly Zholobov was the 35th Soviet cosmonaut.

Colonel Vyacheslav Dmitrievich Zudov

Vyacheslav Zudov was born on 8 January 1942, in the town of Bor, near Gorky, on the Volga. In 1959, he enrolled in the Higher Military Aviation School in Balashov, and graduated in 1963. He subsequently served in military transport aviation, and logged nearly 1200 hours in the air, as well as becoming a highly qualified parachutist. He joined the CPSU in 1963.

He was selected to join the cosmonaut team in October/November 1965, at the age of 23, and served as CapCom for Soyuz 1 in 1967, Soyuz 3 in 1968, and the Soyuz 6, 7, and 8 missions in 1969.

Zudov has been linked with Salyut 3, and it is believed that his first assignment was as the commander of the support crew for Soyuz 14 and Soyuz 15, launched on 3 July and 26 August 1974, respectively. (see section entitled 'Salyut 3 Crewing' for more details.)

His next assignment was as the back-up to Boris Volynov, the commander of Soyuz 21, launched on 6 July 1976, and his first spaceflight came as commander of Soyuz 23, launched on 14 October 1976. The mission called for a docking with the Salyut 5 space-station, but this was aborted when an approach system failure occurred. Zudov and Valery Rozhdestvensky were forced to make an emergency return to Earth, splashing down in Lake Tengiz in the middle of a snowstorm, after a flight lasting only 2 days and 7 minutes. The two men had been formed as a crew two years earlier.

Upon his return, Zudov was awarded the title of Hero of the Soviet Union and also received the Order of Lenin. He is married, and he and his wife, Nina, an engineer at Star Town, have two daughters, Lena and Natasha. During training, he was said to have shown great knowledge, technical erudition, and the ability to make quick decisions. In 1979, he graduated from the Gagarin Air Force Academy while still working at the Yuri Gagarin Cosmonaut Training Centre. He is involved in the Soviet-Ethiopia Friendship Society.

Vyacheslav Zudov was the 37th Soviet cosmonaut

Unconfirmed Cosmonauts

Introduction

Over the years, there have been a number of occasions where Soviet reports have been released naming various cosmonaut-trainees. None have yet flown. This section gives brief details of the men involved:

1: Asanin and Obraztsov

In January 1966, the *Novosti* Press Agency reported an article written shortly before in *'Izvestia'* by a Major Khorobrykh, a military correspondent, who had accompanied a Colonel Asanin and a Lieutenant-Colonel Obraztsov, the flight-surgeon, on a month-long simulation in a 'spaceship cabin' at Star Town.

During the simulation, the two men, described as trainee-cosmonauts, carried out experiments and generally simulated a month-long spaceflight, before 'returning to Earth.'

2: Yuri Vavkin, Ivan Korniev, Rostislav Bogdashevsky.

In 1968, an American NBC TV crew visited the Soviet Union to make a film report on the Soviet space programme. They were supplied with film of cosmonauts training, and were told the names of three of the four men involved. Nothing more was heard of them until 1974, when Bogdashevsky was named in the Soviet *'Litereturnaya Gazata'* as a doctor who took part in the training of cosmonauts. Whether he was also a cosmonaut was not stated. Korniev was reported to be a Major in the Soviet Air Force, having been selected to join the cosmonaut team in 1968.

3: Anatoli Zykov

In the January 1970 issue of the Soviet monthly digest *'Sputnik'*, an article appeared entitled 'No time to fly to the Stars.' It concerned the death of a young engineer named Zykov in a glider crash. He worked in a rocket design bureau, and, it was stated, he applied to join the cosmonaut team on 11 June 1966. He was accepted, only to be rejected later over a query on his height – he was too tall. As height must obviously be a very early disqualifying factor, it is puzzling that Zykov should be accepted and then rejected on this account. The article says that he was accepted, but then goes on to stress that he was **not** a cosmonaut.

4: Vinogradov and Voronov.

On 9 March 1972, the *Agence France Presse* reported that the Soviet Union was preparing to launch a new Salyut, to be manned by two Soyuz craft on a relay basis. The two two-man crews would be made up of Sevastyanov and Kubasov together with two new cosmonauts, Vinogradov and Voronov. Nothing came of the report. However it stated that Popovich was a back-up for the two missions, and both he and Sevastyanov subsequently made Salyut flights. Kubasov is also known to have trained for a Salyut mission. In addition, American ASTP astronauts have reported meeting cosmonauts Vinogradov and Voronov while training at Star Town. Therefore, it would seem that these two 'cosmonauts' are real enough.

5: Barsukov and Bondarev

In May 1977, *Aviatsiya i Kosmonavtika* published a photograph showing two space-suited men, named as 'officers S. Barsukov and K. Bondarev' in a Soyuz trainer. As the photograph appeared in connection with an article about Salyut 3 and Salyut 5, and as it was an all-military crew, it has been speculated that Barsukov and Bondarev may have been a support crew on Salyut 5.

6: Anatoli Antosechin

In late 1984, at an exhibition in Australia, a Soviet scientist named Anatoli Antosechin, from the Institute of Molecular Genetics, said that he was about to start cosmonaut training. It is therefore thought that he will carry out a mission onboard the Mir space-station.

Part 3: Remaining Questions

Part 3: Remaining Questions

Introduction

As stated at the beginning of this book, the Soviet manned space programme is full of mysteries and there are many questions which remain unanswered concerning crew assignments, composition of selection groups, and even the purpose of some of the missions flown. I have attempted to answer some of these questions, but my views must remain speculative in the absence of any firm evidence.

I am sure that there will be readers of this book who have that evidence, or who have additional information which could help fill some of the gaps. If there are, I should be grateful if they would contact me care of the publisher. When writing, it is **essential** to enclose a photocopy of the original source document containing the information, together with the exact title and date of publication. Where appropriate, an English translation should be included.

Questions

Some of the questions which need to be answered are:

1: Soyuz 1 – why was Komarov EVA trained for Soyuz 1, and did the spacecraft have an orbital module?

2: Voronov and Vinogradov – what has happened to them?

3: Cosmos 557 – did Filipchenko/Rukavishnikov, Gubarev/Grechko, and Lazarev/Makarov train for flights to this Salyut attempt?

4: Soyuz 14/15 – were Zudov and Rozhdestvensky the support crew?

5: Soyuz 17 – it is almost certain that Klimuk and Sevastyanov were the back-ups – but where is the official confirmation?

6: Soyuz 40 – why was Bykovsky disqualified, and why did Khrunov decline the mission?

7: Jankowski – he was chosen as the prime Pole – why was he switched to the back-up position?

8: Intercosmos cosmonauts – we know the two finalists, but who were the other short-listed Cuban, Czechoslovakian, Hungarian, Mongolian, Indian, Romanian, Syrian and Vietnamese candidates?

9: 1960 cosmonauts – we now know a little about those men who were dropped from the cosmonaut team, but where is the fine detail?

10: 1963 women – we know about Tereshkova and Tortchillova – what about Irina and the unnamed fourth woman? Who were they?

11: Salyut 1 – who is the ninth man in the training pictures? What happened to him?

12: Soyuz 2 – who was the back-up commander – Nikolayev or Belyayev?

13: Illarionov – what has happened to him?

14: Voskhod 1 – why were two groups of 4 men originally chosen?

15: Soyuz 10/11/12 – who **were** the original prime and back-up crews?

16: 1963 selection – 8 still haven't been named. Who were they and what has happened to them? And which one was the back-up to Patsayev on Soyuz 11?

17: 1965 selection – 14 still haven't been named. Who were they and what has happened to them?

18: 1980 selection – who are the other 3 women? And why hasn't Irina flown yet?

19: 1982 selection – it is believed 6 more women were selected. Is this correct, and if so, who are they?

20: Andreyev – what has happened to him?

21: Soyuz 13 – why haven't the back-ups been announced? Was Lebedev originally the back-up flight-engineer?

22: Soyuz 18 – who was the back-up flight-engineer and what has happened to him?

23: Zond – why was Belyayev's circum-lunar flight cancelled, and who was his back-up – Leonov or Bykovsky?

24: Zond flight-engineers – were Kubasov and Sevastyanov really given Zond assignments?

25: Callsigns – many cosmonauts have never commanded a mission – what were their individual Callsigns?

26: CapCom assignments – there are many gaps – where are the missing details?

27: Belyayev – what was the real cause of death?

28: Dediu and Magyari – what are their dates of birth?

29: Soyuz 24 – who was the back-up flight-engineer and what has happened to him?

30: Friendship Societies – there are many gaps – where are the missing details?

31: Missing biographical information – children:-

Cosmonaut	son	born	daughter	born	Notes
Aksyonov	Valery	?			
	Sergei	?			
Alexandrov (SU)	Sergei	?			
Andreyev	Dmitri	?	Tatyana	?	
Artyukhin	Vladimir	?			
	Sergei	?			
Beregovoi	Viktor	?	Lyudmilla	?	
Berezovoi	?	?	Tatiana	14.7.74	has a son
Demin	Sergei	?	Natalya	?	
Dzhanibekov			Inna	1969	
			Olga	?	
Feoktistov	Andrei	1962	Galina	?	Has one
	Kostya	?			other
Filipchenko	Alexander	?			
	Igor	?			
Ganzorig	?	?			
	?	?			2 sons
Gorbatko			Irina	?	
			Marina	?	
Grechko	Alexei	?			
	Mikhail	?			
Gubarev	Vladimir	?	Olga	?	
Gurragcha	?	?			1 son
Hermaszewski	Miroslaw	?	Emilia	?	
Ivanchenkov			Tanya	?	
Ivanov			Annie	?	
Jankowski			?	?	1 dtr
Khrunov	Valery	?			
Klimuk	Mikhail	?			
Kollner					2 children

Cosmonaut	son	born	daughter	born	Notes
Kovalyonok	Volodya	?	Inessa	?	
Kubasov	Dimitry	?	Ekaterina	?	
Lazarev	Alexander	?			
Lebedev	Vitaly	?			
Leonov			Viktoria Oksana	1963 ?	
Liem			?	?	1 dtr
Lopez-Falcon					1 child
Lyakhov	Yuri	?	?	?	has a dtr.
Magyari			Greta	?	
Makarov	Leonid Kostya	? ?			
Malhotra	Rohid	?	Rakhi	?	
Malyshev	?	?	?	?	1 dtr 1 son
Pelčak	Oldrich Milos	? ?			
Popov	Alexei	?	Yelena	?	
Popovich			Oksana Natasha	? ?	
Prunariu	? ?	? ?			2 sons
Rozhdestvensky			Tanya	?	
Ryumin	Vadim	?	Vika	?	
Sarafanov	Alexei	?	Katya	?	
Savinykh			Valentina	?	
Serebrov	Kirill	?			has one other
Sevastyanov			Natasha	?	
Sharma	Kapil	?	Mansi	?	

Shonin	Andrei	?	Nina	?	
Strekalov			Tatyana	?	
			?	?	two dtrs
Tamayo-Mendez	Arnaldo	?			
	Orlando	?			
Titov,V					Children ?
Tuan			?	?	1 dtr
Vasyutin					2 children
Volk					1 child
Volkov, A	Dima	?			
	Sergei	?			
Volynov	Andrei	?	Tanya	?	
Zholobov			Yelena	?	
Zudov			Lena	?	
			Natasha	?	

32: Missing biographical information – wives

Cosmonaut	**wife**	**Notes**
Atkov	?	Any children ?
Chrétien	?	
Ganzorig	?	
Glazkov	?	Any children ?
Illarionov	?	Is he married ?
Jankowski	?	
Volk	?	

Post Script

Postscript

Soyuz T-15

Launch date	:	13 March 1986.
Crew	:	Leonid Kizim and Vladimir Solovyov.
Back-up Crew	:	Unnamed (rookie ?) and Alexander Alexandrov.
Mission	:	Docked with the Mir space-station.
Callsign	:	Mayak

Progress 25

Launch date	:	19 March 1986.
Mission	:	Docked with the Mir space-station.

New Intercosmos Mission

In March 1986, France and the Soviet Union reached an agreement on a second Soviet-French spaceflight. To be launched in 1988, the mission is expected to use the Mir space-station, and will be a long-duration one of about 30 days, during which an EVA may be conducted by the French crewmember.

Three French cosmonauts – Jean-Loup Chrétien, Jean Francois Clervoy and Michel Tognini – were due to undergo medical tests at the Yuri Gagarin Cosmonaut Training Centre in July 1986. The two finalists were then due to begin training in September/October 1986.

Cosmonaut Death Revealed

In April 1986, *Izvestia* admitted that Valentin Bondarenko – previously identified only as Valentin Junior – died in a fire in a pressure chamber on 23 March 1961. This was the first confirmation of a Soviet training casualty, and it is notable that it had been kept quiet for 25 years. The accident – which was apparently Bondarenko's own fault – came at the end of a 10 day isolation test. He was the youngest member of the 1960 selection and was aged only 24 at the time of his death. He was a Ukrainian, and was married with a young son, Alexander.

1960 Cosmonauts Identified

In an article in *Izvestia* marking the 25th anniversary of Gagarin's flight, the eight missing men from the first selection were finally identified. The details, kindly translated and supplied by Rex Hall, FBIS, were as follows:

Ivan Nikolaievich Anikeyev
Valentin Vasilyevich Bondarenko – Valentin Junior
Valentin Ignatevich Filatev – Valentin 'Dyed'
Anatoli Yakovlevich Kartashov
Grigori Grigorievich Nelyubov
Mars Zakirovich Rafikov

Valentin Stepanovich Varlamov
Dmitri Alexeivich Zaikin

It was confirmed that Zaikin had served as back-up on Voskhod 2, and it was revealed that Nelyubov – dismissed from the cosmonaut team for disciplinary reasons – committed suicide by jumping in front of a train on 18 February 1966.

ORDER FORM

For further copies of this book, please send the following order form with your remittance to:

GRH Publications,
36, Bury Hill,
Melton 1001,
WOODBRIDGE
Suffolk
IP12 1LF
ENGLAND

Please send me .... copies of 'THE SOVIET COSMONAUT TEAM' by Gordon R. Hooper.

A cheque for £11.95 (£10.95 + £1.00 postage and packing) per copy is enclosed.

Name ____

Address ____

____

____

____

____

Overseas Orders

Overseas customers should pay by International Money Order or Sterling Bank Draft as follows:

Europe	—	£10.95 + £3.05 postage and packing.
United States and other countries outside Europe	—	\$17.00 + \$4.00 postage and packing – Surface. \$17.00 + \$12.00 postage and packing – Air Mail.

ORDER FORM

For further copies of this book, please send the following order form with your remittance to:

GRH Publications,
36, Bury Hill,
Melton 1001,
WOODBRIDGE
Suffolk
IP12 1LF
ENGLAND

Please send me .... copies of 'THE SOVIET COSMONAUT TEAM' by Gordon R. Hooper.

A cheque for £11.95 (£10.95 + £1.00 postage and packing) per copy is enclosed.

Name ______________________________

Address ______________________________

Overseas Orders

Overseas customers should pay by International Money Order or Sterling Bank Draft as follows:

Europe	—	£10.95 + £3.05 postage and packing.
United States and other countries outside Europe	—	\$17.00 + \$4.00 postage and packing – Surface. \$17.00 + \$12.00 postage and packing – Air Mail.